NATURE CONTESTED

NATURE CONTESTED

Environmental History in Scotland
and
Northern England since 1600

T. C. SMOUT

EDINBURGH
University Press

What would the world be, once bereft
Of wet and of wildness? Let them be left,
O let them be left, wildness and wet;
Long live the weeds and the wilderness yet.

Gerard Manley Hopkins (1844–89), *Inversnaid*

© T. C. Smout, 2000

Transferred to Digital Print 2009

Edinburgh University Press Ltd
22 George Square, Edinburgh

Typeset in Bulmer
by Pioneer Associates, Perthshire, and
Printed and bound in Great Britain by
CPI Antony Rowe, Chippenham and Eastbourne

A CIP Record for this book is available from the
British Library

ISBN 0 7486 1410 9 (hardback)
ISBN 0 7486 1411 7 (paperback)

The right of T. C. Smout
to be identified as author of this work
has been asserted in accordance with
the Copyright, Designs and Patents Act 1988.

CONTENTS

ACKNOWLEDGEMENTS

My first thanks must be to the Ford's electors of the University of Oxford who invited me to give the lectures on which this book is based, and to St Catharine's College who elected me to a visiting Christensen Fellowship for the duration of my stay in the city. The University of York honoured me with a Visiting Professorship in 1998–9 which enabled me to improve my knowledge of the environmental history of the north of England, assisted by their friendly historians and economic historians. My own University of St Andrews was patient in my endless search for library materials. To my wife Anne-Marie I owe boundless support, and to Margaret Richards equally boundless patience in fielding one draft after another. David Jenkins was a kind and severe critic of the first draft: I hope he finds it improved, but I cannot promise to please him. Different parts were read and commented upon by Donald Davidson, Jeff Maxwell, John Sheail, Robert Lambert and Roger Crofts. I owe many debts of conversation and information to other friends and colleagues in Scottish Natural Heritage – Magnus Magnusson, Michael Usher, Des Thompson, John Mackay, John Thomson, Alan Macdonald, Chris Badenoch and Dick Balharry. To Chris Lowell I owe endless stimulation as we drove round Scotland preparing broadcasts for the Radio Scotland 'Battle for the Land' series. I am afraid he may recognise some of his ideas and observations unattributed. One person we contacted at that time was Catherine Benton of West of Scotland Water, and I owe a debt to her and to the company for the opportunity to use papers in their offices at Milngavie. The archivist at the National Trust for Scotland was similarly helpful: I would like to spend more time in that fine archive. John and Philippa Grant allowed me into their attics at Rothiemurchus and plied me with much kind hospitality whenever I emerged. The Duke of Buccleuch generously allowed access to the Drumlanrig game books. Roland Paxton and Michael Chrimes gave me advice on where to look for water engineers. Richard Smout found a reference to sparrows I could not resist using, even if it does relate to the Isle of Wight. Dr Bill Bourne shared his expertise on medieval herons. Derek Langslow of English Nature and Jeremy Greenwood of the British Trust for Ornithology provided me with material for two of the figures in Chapter 3. There are many other friends and scholars to whom I owe debts – Jim and José Harris, who drew my attention to the opinions of J. S. Mill on property and the environment; Sylvia Price for allowing me to prowl her books and for so much encouragement; Seumas Grant for help with the Gaelic poetic tradition; Hugh Ingram for lending me early books on bogs; Donald Woodward for inspiration on manures; Lorna Scammell for helpful discussions; Judith Garritt and Richard Tipping for permission to cite their unpublished papers; Judith Tsouvalis-Gerber for insights on the Forestry Commission; Reay Clarke for his expert knowledge of Sutherland grazing. John Davey and his colleagues at Edinburgh University Press have been extremely helpful and supportive in the uneasy process of turning lectures into a book. This text has been cooking in my head for so many years that I fear I have overlooked many others whose wisdom and help went into the pot. But I thank them all.

Finally there came the pleasant task of choosing the illustrations. They are acknowledged individually on the captions, but I received great kindness and help from staff in St Andrews University Library, Dundee University Library, the British Library, the National Library of Scotland map room, the Royal Commission on the Ancient and Historical Monuments of Scotland, Cambridge University Air Photo Collection, Cambridge University Zoology Department, the Scottish National Portrait Gallery, Scottish Natural Heritage, English Nature, the Royal Society for the Protection of Birds, West of Scotland Water and the Ramblers' Association. The Duke of Buccleuch, Michael Dower, Roy Dennis, Donald Davidson, Neil McIntyre and Laurie Campbell all generously allowed me to use pictures in their own possession. Had there been room for as many again, I could not have resisted them.

The endpiece at the conclusion of Chapter 6 is from N. E. Hickin, *The Natural History of an English Forest* (Shrewsbury, 1978). The remainder are from the work of Thomas Bewick and from *Tallis's Scripture History for Youth* (ed. W. I. Bicknell, London, mid-nineteenth-century). I acknowledge the subtle genius of them all.

Chris Smout
Centre for Environmental History and Policy,
Universities of St Andrews and Stirling.

LIST OF ILLUSTRATIONS

LIST OF FIGURES

INTRODUCTION

The occasion of this book was an invitation to give the Ford Lectures on British History at the University of Oxford in January and February 1999. I chose as my title, 'Use and delight: environmental history in northern Britain since 1600', and the text here follows the structure, argument and style of what I said on those six consecutive Friday evenings. I have, however, sometimes changed the wording to make it appropriate for written form, and added new material, either to expand the argument or to enlarge the range of examples, in order to make the exposition clearer or better supported.

The roots of my interest in the subject run further back and deeper. Firstly, as an economic and social historian of Scotland I have long felt that the history of land use has been relatively neglected for at least four decades. Alexander Fenton's explorations in rural ethnology, invaluable though they are, do not make up for the absence of a comprehensive agrarian history of the kind that Joan Thirsk and her colleagues have been providing for England and Wales.[1] On the other hand, it was not in my competence (or responding to my deeper instincts) to attempt a true agricultural history, and this enormously worthwhile undertaking still awaits its scholars. The new concept of environmental history, however, offered some alternative in tune with my proclivities. Since the 1970s it has been put to good use by the historians of Australia, the United States, Africa and India, and is now attempted in different ways in Britain, the Netherlands and Scandinavia. To use it (essentially in essay form) would allow me to consider fresh perspectives in land use, and it was one which would also allow me to stray south over the Border, since environment scarcely recognises a political frontier. Because this book springs from this specific root of interest, it is not an environmental history *of* Scotland and northern England, but one about environmental history *in* that zone of Britain. It is focused on the countryside, and only incidentally involves the town. It has more to say about landscape, land utilisation, nature conservation, biodiversity and access than about urban pollution, transport or the-country-in-the-town.

A second root comes from my experience when, in the mid-1980s, I was appointed to the Scottish Advisory Committee of the Nature Conservancy

1

Council, meeting in Edinburgh, and later to the Board of its successor body, Scottish Natural Heritage, for whom I served as Deputy Chairman from 1992 to 1997. From the day that I first sat down at the long table in 12 Hope Terrace, I was struck by the passions unleashed by the difficult cases that came before us – here was hotly contested ground – and by the essentially historical nature of many of the problems. If a wood needed saving, it was because it had a history in which human beings had once played a central part, and because today other human beings (with their own histories as foresters) wished to play a different part. When a Highland sea loch was proposed as a marine nature reserve, the anger which this aroused had to do with ancient concepts of usufruct and property, opposed to more recent (but also historical) concepts of heritage and public interest. Yet both sides in such conflicts tended to see them as problems with only contemporary and immediate significance, nature versus the developers, jobs versus birds, right versus wrong.

Yet people are part of nature. To say this is not to abandon the convenient language of talking about 'people' on the one hand and 'nature' on the other: in this sense, a habitat that is 'more natural' is simply one less influenced by human activity. But our own incontestable naturalness confines us to personal mortality, evolutionary fragility and total dependence on the natural systems of the globe.

Once we recognise our character, we can see choices more clearly. We can, perhaps, 'respect nature' by deciding to use it lightly and intervene as little as possible – as would be ideal on the fragile high tops of the Cairngorms. But, in pursuit of our own legitimate human economic and social aims, we can also intervene much more basically, and this need not be a disaster: we can find ways to enrich biodiversity by multipurpose uses of a wood, by organising beetle banks in arable fields, or maintaining certain levels of grazing and burning on the moors, by digging gravel pits for industry and creating important wetlands afterwards. These things work with nature. Alternatively we can impoverish a countryside by misuse of pesticide, by developing genetically modified crops of a type that would allow the total destruction of wild flora and invertebrates on a farm, or by promoting (as we currently do) absurd levels of overgrazing by sheep and deer. These things work against nature, belittle and abuse it, and some (precisely because we are inescapably part of nature) could rebound with terrible consequences to ourselves. All courses of action have economic and other costs; in those that work against nature, they are more often hidden and loaded onto the future.

Not to appreciate that people are truly part of nature has helped to drive conservation and developers into confrontation – *we* must stop *them* doing this to nature: *they* must not stop *us* doing that for people. For a time, the language of international agreement at Rio, involving the concepts of respecting 'sustainability' and 'biodiversity', seemed to offer the prospect of better understanding. 'Sustainability' implies the obligation to use natural resources, including the beauty of the earth itself, so that the prosperity and happiness of future generations is not compromised. 'Biodiversity' comprises the plenitude of life and ecosystems.

Yet, to judge from the conflicts that still periodically engulf the countryside, it is doubtful whether confrontation has really receded far. The instincts of conservationists are still deeply distrustful, and in some cases knee-jerk aggressive, though few of them would now regard rural economic development as an undesirable aim in itself, or deny the democratic need for local communities to be involved in reaching a consensus about the use of natural resources. On the other side of the fence, too many developers and politicians, and even some farmers and landowners, give the impression of understanding sustainability and biodiversity only as catchwords, as a chance for spin. Sustainability seems in the mouth of cabinet ministers to mean no more than economic growth without bumps: biodiversity is in some danger of slipping out of use, as if it were too obscure or unsuitable for the voter.

Too often those with economic and political power still see nature as an enemy to struggle against and to be compelled to yield up wealth and employment. They give little consideration to what may happen to posterity, when they have been voted out of office and their retirement homes have crumbled into dust. And they have too little understanding that we delight in our environment as well as utilise it, that we are spiritual as well as material beings. For the human species, considering its unparalleled power over the rest of nature, the duty of care should be seen both as an imperative moral responsibility and enlightened self-interest. Nothing less does our full humanity justice.

There is a preacher in me, which the historian in me needs to distrust. Those years in the NCC and SNH confirmed a desire to write what the Scottish Enlightenment would have called 'useful history'. To admit as much is almost tantamount to disqualifying oneself as a serious academic, for surely useful history becomes 'Whig history', vitiated by a desire to justify the attitudes of the present by mining the past for facts that suit a predetermined case.[2] It could easily slip into being so, yet few in the historical profession would want to admit to writing totally useless history. We want to write available history: within our limits, to counteract our natural bias and to write a history neither to justify nor direct the present, but to inform it. As the following pages will show, it is far from being the case that our species in the past has always benevolently interacted with the environment: waste, selfishness and short-sightedness have been part of the human story as far back as we can trace it. There is no golden age, pre-capitalist, pre-Christian or prehistoric, to return to. But there have been other parts of the story, too, and if we listen to the past we would have to be deaf not to learn something.

As long ago as 1848, John Stuart Mill, the most astute, challenging and widely read political economist of his day, contemplating the pros and cons of the 'stationary state' – an economy that had ceased to grow – then began to ask what would be the costs of growth in a world over-used and over-populated. In the following arresting passage he identified the need for space and for wildlife as part of human happiness:

> It is not good for man to be kept perforce at all times in the presence of his species. A world from which solitude is extirpated is a very poor ideal. Solitude, in the sense of being often alone, is essential to any depth of meditation or of character; and solitude in the presence of natural beauty and grandeur, is the cradle of thoughts and aspirations which are not only good for the individual, but which society could ill do without. Nor is there much satisfaction in contemplating the world with nothing left to the spontaneous activity of nature; with every rood of land brought into cultivation, which is capable of growing food for human beings; every flowery waste of natural pasture ploughed up, all quadrupeds or birds which are not domesticated for man's use exterminated as rivals for food, every hedgerow or superfluous tree rooted out, and scarcely a place left where a wild shrub or flower could grow without being eradicated as a weed in the name of improved agriculture. If the earth must lose that great portion of its pleasantness which it owes to things that the unlimited increase of wealth and population would extirpate from it, for the mere purpose of enabling it to support a larger, but not a better or a happier population, I sincerely hope, for the sake of posterity, that they will be content to be stationary, long before necessity compels them to it.[3]

It is impossible to read these words without feeling that the scenario he dreaded is now taking shape before our eyes, and that his warning went unheeded and forgotten because it did not appeal to practical men of the world with access to the capital to change it. This book does not deal with global issues or ecodoom, but it does address concerns that Mill would have recognised at home, in the tension between use and delight. How have we treated the countryside of Scotland and northern England since 1600, what attitudes have we brought to nature, and what pickles have we landed ourselves in as a result? To understand how we arrived at our current local predicaments might, just possibly, help to clear our heads about directions to take in the future.

CHAPTER 1

USE AND DELIGHT:
ATTITUDES TO NATURE
SINCE 1600

We live within the natural world and we derive delight from it. The human attitude to the environment has differed over time and place as we construct nature differently from generation to generation, yet it has always been shaped by the twin considerations of use and delight, certainly in the West since before Horace used the phrase about his garden in the first century before Christ. Now we have the problem that our use has become so intensive that not only is our delight threatened, but possibly our survival. This may not be the first environmental crisis in history, but it is the first time that we have threatened the globe rather than merely ourselves and a few ambient species on a local scale. Nor are we necessarily incapable of surmounting the crisis in which we are placed, or with which we are likely to be faced in the future in much more acute form: but our ultimate success is far from being a foregone conclusion.

The point of this book is to take a part of Great Britain, roughly the northern half, and a period of the last four hundred years, and on this theatre of space and time to ask how the tension between use and delight has worked out. In those centuries, science and industrialisation have so transformed our relationship with nature that at the close of the second millennium we can even contemplate the possibility of the end of the world, not as an act of divine judgement as our forebears would have done in 1600, but as the final act of human folly.

Let me first explain the geographical theatre. By northern Britain I mean a country rather more like Wales or Ireland than southern England. It begins with the Pennines. Doncaster, Derby and Manchester are frontier towns. It ends, terrestrially, with Muckle Flugga, the lighthouse off the north end of Unst in Shetland. Edinburgh, half-way between Unst and London, is well inside the southern half of this northern territory (see Figure 1.1). If my centre of gravity is Scottish this reflects not only my professional expertise but the logic of my division of Britain.

What biogeographical sense it would have made if only the medieval dynastic struggle had ended differently, with the Capetian Kings of France ruling England from the Garonne to the Humber, and the House of

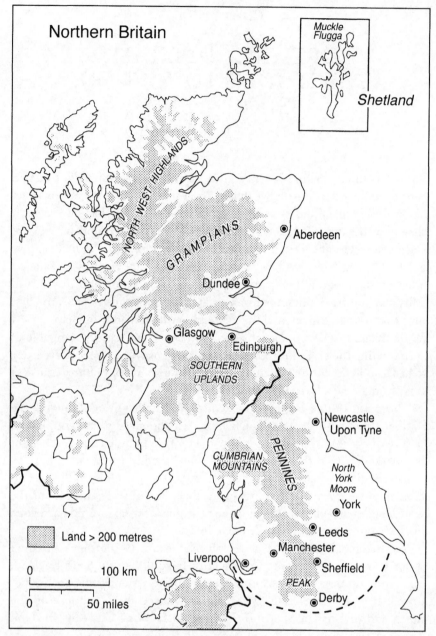

Figure 1.1 Northern Britain.

Stewart north from there. The Plantagenets could have contented themselves with their rich patrimony of Gascony. Whereas the south would have
had the mega service-cities Paris and London, the heavy clays and chalk,

*Northern Britain: Suilven and Canisp in Wester Ross – a quartzite
cap over Torridonian sandstone.
Cambridge University Collection of Air Photographs.
© British Crown Copyright/MOD.*

the marshes and the majestic woods of pedunculate oak and beech of southern England and northern France, the north would have had most of the wild land of Britain – wild but not unaltered: the greatest stretches of moor and mountain; more than 95 per cent, by volume, of the standing fresh water (Loch Ness alone contains more than all the lakes and reservoirs in England and Wales); woods of quite a different character, dominated by sessile oak, Scots pine and birch, richest in bryophytes where most drenched by rain. It would contain most of the coal reserves from much earlier forests and therefore most of great cities of industrialisation, Manchester, Leeds, Sheffield, Newcastle, Glasgow. In the opinion of Sir Martin Holdgate, it would contain most of the 'habitats, ecosystems and species which most distinguish this country in a European context' – areas of patterned mire (of which Britain has about 13 per cent of the globe's resource), oceanic montane habitats like the western Highlands, areas of limestone karse with ancient erosion features like those of northern

England, seabird colonies and other sea coast habitats with a high-oceanic climate like those of St Kilda, the Hebrides and the Northern Isles.[1] The length of its sea coast and therefore the pervasive influence of the sea is extraordinary: Scotland alone has more coastline than the eastern United States.

But of course northern Britain in the last four centuries has also had many things in common with southern England, and indeed with other low-lying countries of north-west Europe such as the Low Countries and Denmark: important if smaller service centres like Edinburgh and York, or now like Glasgow, Aberdeen and Manchester; stretches of prime arable country as in the Vale of York, Northumberland, Lothian, parts of Fife and Angus. At the present time it contains 19.5 million people, or rather over a third of the population of Great Britain. In 1700 it contained 2.2 million, or rather under a third of the population. Within that, however, Scotland has lost population relative to the north of England. Now only about a quarter of the population of northern Britain live in Scotland: in 1700 it was about half. The biggest relative decline has been in the Scottish Highlands, but everywhere the country has lost to the city and the town. Northern Britain covers about half of the land surface of Great Britain but contains well over half of the land designated under the 1981 Wildlife and Countryside Act as Sites of Special Scientific Interest. In 1998 there were 1,441 SSSIs in Scotland alone, covering 11.6 per cent of the total land area.

In this chapter, I want to look harder at notions of use and delight in our encounter with the natural world. The inhabitants of northern Britain, not least those of the north English counties, had their own distinctive view of their environment, and in some respects importantly changed the dismissive opinions of the southern, metropolitan English. In the next chapter, I will address the fate of the woods, that most emotive and multi-purpose of all terrestrial resources. In the following four, I will examine soil, water, the hills and the modern quarrel over the countryside, which is certainly not confined to the north but affects the south just as much. By the end, I hope to have presented on my theatre a dramatic tale that tells us a fair amount not only of northern Britain, but something about the globe and the European west as a whole over the last four hundred years.

So what of use and delight? They are two sides of the coin in our attitude to Nature and have always been so, though what we have used and what we have delighted in have not always been the same. People in northern Britain around 1700 regarded the natural world in which they lived essentially as God's *fait accompli*, part of the Divine handout that followed the expulsion from Eden: it was here you lived by the sweat of your brow, here you courted, mated, gave birth and died. For the Yorkshire-born

Northern Britain: Clawthorpe Fell National Nature Reserve,
Cumbria – limestone pavement.
Peter Wakely, English Nature.

divine Thomas Burnet, whose *Sacred Theory of the Earth* attracted vast admiration when it was published in 1681, the earth had once been a near-perfect sphere or, as he called it, a 'Mundane egg' – 'all one smooth continent, one continued surface . . . it had the beauty of youth and blooming nature, fresh and fruitful', without scar or fracture, rock, mountain or

cavern. Here Adam and Eve had lived amid a harmonious nature, sustained
by a layer of life-giving substance or 'terrestrial liquors'. The Fall of Man
had broken this harmony, and the Flood had wrecked the globe, collapsing
so that 'the frame of the Earth broke and fell down into the Great Abysse',
releasing subterranean fires and waters and throwing up great mountains.
Bizarre though this theory may now appear, Burnet, who was a Fellow of
the Royal Society, was attempting to square theology with Restoration
science, explaining natural change by natural laws and so reducing the
need for incessant miraculous intervention. Still, in his Christian tradition
we lived amid the 'ruins of a broken world', an imperfection brought about
by human disobedience and sin.[2]

But if it was theologically a vale of tears, to many it was manifestly a
material vale of delight. I can trace no fear of the natural world around them
in native-born commentators of northern Britain. It was the visitors to
whom it was famously alarming. The dismay of that stout Londoner,
Daniel Defoe, on entering Westmorland was palpable:

> A country eminent only for being the wildest, most barren and frightful of
> any that I have passed over in England . . . seeing nothing around me, in
> many places, but unpassable hills, whose tops, covered with snow, seemed
> to tell us all the pleasant part of England was at an end.[3]

Defoe was more at home on Cheapside than on Ambleside. No less of a
southerner was Edward Burt, stationed as a soldier in the Highlands
between the Jacobite rebellions in a depressing world of mountains 'gloomy
brown drawing upon a dirty purple and most disagreeable when the heath
is in bloom'.[4] And Dr Johnson, on his visit to Scotland in 1773, spoke of a
land almost totally covered with useless, dark and stunted heath, matter
incapable of form: 'an age accustomed to flowery pastures and waving
harvests is astonished and repelled by this wide extent of hopeless steri-
lity'.[5] Similarly, the landscape painter John Constable, accustomed indeed
to the flowery pastures and waving harvests of Suffolk, confessed after a
visit to Wordsworth that 'the solitude of mountains oppressed his spirits'.[6]

Far more relevant, and almost always overlooked, is the opinion of the
mountain world by those who lived there. Let me set against Defoe and
the others this anonymous seventeenth-century account of an area much
wilder and more remote than the English Lakes. It lies on the northern
shores of Sutherland:

> Now let us speak something of Strathnaver . . . [It] is a countrey full of
> bestial and cattel, fitter for pasturage and store than for corns, by reason
> there is little manured land there . . . The principal commodities of

Strathnaver are cattel and fishing, not only salmon, whereof they have great store, but also they have such abundance of all other kinds of fishes in the ocean, that they apprehend great numbers of all sorts at their very doors, yea in the winter season among the rocks without much trouble . . . There are diverse rivers in Strathnaver, wherein they do take good store of salmon . . . The Countrey of Strathnaver is full of red deer and roes, pleasant for hunting in the summer season. It is full of great mountains and wilderness, yet very good for pasture. It is stored with all kinds of wild fowl . . . There are diverse lakes or lochs in Strathnaver, where of the chiefest is Loch Naver, full of good fishes. In Loch Loyal there is an island which is a pleasant habitation in the summer season. Macky hath also a summer dwelling in an island within Loch Stalk in the Diri More. Thus much of Strathnaver.[7]

This was an entirely characteristic topographical account of Scotland as written at any time between the late sixteenth and early eighteenth centuries. Our author does not attribute any special aesthetic or scenic value to the wildness of mountain, shore or river, but he is neither repelled nor intimidated by his environment. On the contrary, for him Strathnaver is a delightful place, Virgilian in its pastoral peace, rich in natural resources for use, with excellent hunting grounds for pleasure, and the perfect spot for a second home in summer. *Et in Arcadia ego.*

The same tone, unintimidated, engaged and pleased, comes out in the manuscript notes of Timothy Pont on his draft maps of the far north-west Highlands, *c.* 1590 – 'Item I fand the two kynds of mos wyld berries with ther floures in the head of Korynafairn': 'Item . . . the mightiest and largest salmon in al Scotland ar in this Loch [Loch Shin]': 'Extreem wilderness': 'heir ar black flies in this wood . . . souking mens blood': 'the wood fowel ludge here often on thes ylands'; 'many wolfs in this cuntry'; 'guid corn and fishing'; 'fair lynks and bent'; 'many lusty stuids'; 'excellent hunting place wher are deir to be found all the year long as in a mechtie parck of nature'.[8] Also possibly from Pont's pen, and certainly of similar date, is the first account of the woodlands round Loch Maree:

compas'd about with many fair and tall woods as any in all the west of Scotland, in sum parts with hollyne, in sum places with fair and beautifull fyrrs of 60, 70, 80 foot of good and serviceable timmer for masts and raes, in other places ar great plentie of excellent great oakes, whair may be sawin out planks of four sumtyms five foot broad. All this bounds is compas'd and hem'd in with many hills, but thois beautifull to look on, thair skirts being all adorned with wood even to the brink of the loch for the most part.[9]

Northern Britain: Borrowdale, Cumbria – oakwoods, fells and fields.

In all this, the natural world, however mountainous, forested or remote from the city, is delightful wherever it is useful for economy or pleasure, and the author does not scruple to describe as beautiful, wooded hillsides similar to those that Defoe saw in the English Lakes 'when all the pleasant part of England was at an end'.

A point of clarification is needed here. Burnet, who had seen the Alps and the Appenines on a visit to Rome, considered that such mountains had 'neither form nor beauty nor shape nor order' yet, like the wide sea, they inspired 'great thoughts and passions', casting the mind 'into a pleasing kind of stupor and admiration'. In this he was foreshadowing (and helping to form) the taste of the following century.[10] Most commentators describing the countryside of northern Britain, however, were not particularly enamoured of wilderness for its own sake. Rather, to them, the local hills and glens were inhabited, familiar and friendly places, though they might have wild parts that would, in particular, intimidate strangers. Thus Alexander Pennecuik on Tweeddale, with particular reference to the desolate country round the Devil's Beef Tub at Moffat:

> This country is almost everywhere swelled with hills; which are, for the most part, green, grassy and pleasant, except a ridge of bordering mountains, betwixt Minch-Muir and Henderland, being black, craigie and of a melancholy aspect, with deep and horrid precipices, a wearisome and comfortless piece of way for travellers. The vallies are not big, but generally pleasant to the view, fertile of corn and meadow and excellently well watered.[11]

Real wilderness, 'in the properest sense', was indeed horrendous, but it was not often considered to be in Scotland. In 1700, the Reverend Alexander Shields found himself chaplain to the second expedition of the Company of Scotland attempting to settle Darien on the coast of Panama. To their dismay, they found the first colonists fled and the settlement deserted. He wrote home to his kirk session in St Andrews:

> Instead of the comfortable settlement we expected, nothing left but a vast howling wilderness in the properest sense; here all the circumstances of unpassable woods, vast desolations never frequented by mankind, retired recesses and resting places of tigers, buffaloes, monkeys and other wild beasts, all manner of dangers, distresses and difficulties that can make any place a wilderness.[12]

Shields was to die on the expedition. I am told by an archaeologist who visited the site at Darien that his own aesthetic reaction was not very different, though his fate was better.

From the prosaic topographers, let us turn to poetry, and especially that of the Celtic peoples who lived in closest touch with the natural environment of the north and west. As Derick Thomson and James Hunter have emphasised, there was a tradition in Irish Gaelic verse, as far back as the ninth and tenth centuries, of speaking directly of the natural world in a manner quite exceptional in European literature:

> I have tidings for you: the stag bells; winter pours; summer has gone;
> Wind is high and cold; the sun low; its course is short; the sea runs
>   strongly;
> Bracken is very red; its shape has been hidden; the call of the
>   barnacle-goose has become usual;
> Cold has seized the wings of birds; season of ice: these are my tidings.[13]

Irish and Scottish Gaelic poetry share a common tradition in which sensitivity to the natural world remained a general underlying feature for a millennium, though its tone changed. In the seventeenth century, Scottish Gaelic poets made use of the 'pathetic fallacy' in the praise poetry of the age, the idea that nature laments in sympathy with human emotions.[14] By the eighteenth century this has again changed. Derick Thomson describes a new Gaelic poetry in which nature is an end in itself, 'not a facet of poetry on other themes and, unlike the very early poems on the seasons (in Irish) has a length, scale and detail that are new in the tradition'. He attributes this in part to the impact of James Thomson, the Roxburghshire poet, whose *Seasons* (1726–30), with their unforced observation of nature, and delight in storm and hillside, were also absolutely seminal to English romantic poetry. The Ardnamurchan poet, Mac Mhaighstir Alasdair adapted and transformed this vision into something still more direct so that he became 'almost clinical in his observation and reporting of the natural scene: he does not let any moral reflection intrude, as James Thomson does.' Mac Mhaighstir Alasdair's work in turn had a critical impact on Duncan Ban Macintyre, and a series of other eighteenth-century poets like Rob Donn, Uilleam Ros and Ewen MacLachlan.[15]

Of these, Duncan Ban Macintyre was unquestionably the finest. He was an illiterate gamekeeper whose Gaelic songs were first written down and published in 1768 when he had left his job in the Breadalbane Highlands and found another in the town guard in Edinburgh. Here is a scholarly modern translation of the opening stanza of *The Song to Misty Corrie*:

> The Misty Corrie of the roving young hinds
> is the dearest corrie of verdant ground;
> how lovely, lea-loving, sleek-bright, sappy,
> was every floweret so fragrant to me;
> how shaggy, dark-green, fertile, teeming,
> steep, blooming, full pure, exquisite,
> mellow, dappled, flowery, bonny, rich in sweet grass,
> is the glen of arrow grass, and many a fawn.

Elsewhere in this fine poem he speaks of mountain springs with their 'sombre brow of green-watercress' becoming streams:

> splashing gurgles, seething, not heated,
> but eddying from the depth of smooth cascades,
> each splendid rill is a blue-tressed plait,
> running in torrent and spiral swirls.

He describes the salmon leaping:

*Northern Britain: above Horton in Ribblesdale – a drumlin field,
the refuse of a departing glacier.*
Cambridge University Collection of Air Photographs. Copyright reserved.

> in his martial garb of the blue-grey back,
> with his silvery flashes, with fins and speckles,
> scaly, red-spotted, white-tailed and sleek.

And above all, the gamekeeper comes out in Duncan Ban Macintyre:

> deer will be slung up when blast of powder
> driveth dark-blue lead thick into their pelts;
> the gun is ready, and the whelp is nimble,
> blood-thirsty, forceful, valiant, fierce,
> careering fleetly, bounding briskly,
> stretched to the utmost against a red flier.[16]

The language is like Gerard Manley Hopkins in its compression and intensity in its directness and unsentimentality like John Clare.

The Gaelic poets were in every sense poets of the people, and their language is of a direct delight in nature, not of nature as a vehicle for moral introspection as in the English Wordsworthian mould. The popularity of Duncan Ban Macintyre's songs in Perthshire and Argyll paralleled in some ways the appeal of Robert Burns to the tenants and farm servants of Ayrshire: his view of nature and life, like that of Burns, struck a chord among ordinary unlettered people.

Let me return you to the main argument so far. It is that we should not be misled by the disparaging remarks of outsiders who were unfamiliar and therefore uncomfortable with the natural world of northern Britain into believing that the inhabitants themselves did not delight in it. Yet every age constructs nature in a different way. In the seventeenth century, use and delight were very difficult to separate. By the end of the eighteenth, they occupied different spheres in the mind. By the twentieth, they were in frequent conflict. Let me explain.

It was characteristic of seventeenth-century Scottish topographers that they attempted to describe the world as they saw it. They criticised their predecessors, by implication especially Hector Boece, the first principal of Aberdeen University and highly influential historian of Scotland, writing at the hinge of the fifteenth and sixteenth centuries, as sacrificing the plain truth to a fabulous tale in the medieval manner. In the central decades of the seventeenth century, Sir Robert Gordon of Straloch told his collaborators 'not to have anything in their writings too extravagant or beyond the truth, nor make an elephant of a fly . . . the faithful and full description of our districts remains untouched'.[17]

Gordon, like Sibbald and eventually Sinclair after him, was in the business of gathering what he considered to be the facts about Scotland. Harold Cook has pointed out that the modern sense of a fact as something that had 'really occurred . . . hence a particular truth known by actual observation or authentic testimony, as opposed to what is merely inferred', was new to the seventeenth century. It was consciously Baconian in its empiricism. But the account the topographers gave with their 'facts' described a harmonious but static world: one that would have been familiar to Horace and Virgil, useful and delightful together, but static:

St Mary Loch is in circuit at least six miles, surrounded with pleasant green Hills and Meadows; the hills overspread with flocks of sheep and Catel, the Rocks with herds of Goats, and the Valleys and Meadows with excellent corn and Hay. It is fed with several little Springs and Brooks but chiefly with the Water of Meggit which with a clear stream runs gently down a long Plain and discharges itself prettily in its very bosom.[18]

William Gilpin as Dr Syntax setting out in search of
the Picturesque, by Thomas Rowlandson.

By the time the end of the eighteenth century has been reached, topographical descriptions have become dynamic. Nature was now seen to
provide a launching pad for improvement. So when Sir John Sinclair
published the *Statistical Account of Scotland* in the 1790s, it was full of
observations like this:

Of the land which at present lies waste, a considerable extent may be
brought into a state of cultivation. The meadows, in particular, on the

banks of the Motray, and which the water often overflows, might be made
the most productive ground in the parish. Activity has already converted a
part of these into the most luxuriant corn-fields; but to the improvement of
the whole an obstacle is opposed, which perhaps may not speedily be
removed.[19]

The obstacle in question was four watermills whose proprietors feared
losing their power. We call that change in attitude, some would say with an
irony that is still unconscious, the Enlightenment. The fruits of the
Baconian and Newtonian Revolutions were a resolve not to accept nature
as an unalterable given, a *fait accompli*, but as an enormous unrealised
opportunity, the material from which man could fashion his worldly
improvement if he were daring and knowledgeable enough. In Scotland,
especially, the phrase 'improver' came to be associated with a whole new
attitude to natural resources, one which was searching and critical, looking
constantly for opportunities to change. 'We war with rude nature,' said
Thomas Carlyle, 'and by our resistless engines, come off always victorious;
and loaded with spoils'.[20]

Take the attitude to peat bogs as an example. To Sir Robert Gordon and
his seventeenth-century contributors, a bog was a bounty of a merciful
providence, which provided 'inexhaustible mosses, wherein are digged the
best of peats': seldom is there mention of any possible change in land
use. To the improvers, however, it was a scandal that, as Robert Rennie (a
pioneer in the study of bog formations) said in 1807, 'innumerable millions
of acres lie as a useless waste, nay, a nuisance to these nations', or, as Andrew
Steele put in the opening chapter of a treatise in 1826, bogs were 'immense
deserts . . . a blot upon the beauty and a derision to the agriculture of the
British Isles'. He also commented that 'the only animals found on these
grounds are a few grouse, lizards and serpents'.[21]

So bogs were a challenge, and philanthropic and far-sighted men like
Lord Kames in Stirlingshire reclaimed them and put the poor to work.
William Aiton compared his Herculean efforts to drain the great moss at
Blair Drummond to those of David Dale to create his famous cotton mill at
New Lanark, entirely to the favour of Kames. That 'several hundreds of
ignorant and indolent Highlanders' were 'converted into active, industrious
and virtuous cultivators, and many hundreds of acres of the least possible
value rendered equal to the best land in Scotland are matters of the highest
national interest, to which I can discover no parallel in the cotton mill
colony'.[22]

Scottish improvers have had a mixed press, but the notions of the
Enlightenment certainly became a no-nonsense, let's-get-moving, wholly

materialist and developmental view of the world, which has been the dominant ideology in Europe and America since the eighteenth century, in Japan since the nineteenth century, and in virtually all the nations of the globe in the twentieth.

Yet the concept of improvement in the eighteenth century was by no means entirely consumed with material progress, and therein lay a contradiction. The nobility had long embellished their policies with gardens and parks: as landed incomes rose, more and more of the landed classes could afford to follow the twists of landscaping fashion. 'Improving' could mean improving your estate for pleasure as well as improving it for profit. Fashion began to lead towards an aesthetic appreciation of untamed nature in the garden or on the estate policies, blurring the distinction between nature and art. You could improve things for your own delight by making them wilder.

This confusing idea can be found early in the eighteenth century, even in such classical authors as Pope and Shaftesbury. Addison condemned the 'formal Mockery of princely Gardens' and declared 'the passion in me for things of a *natural* kind'.[23] Partly this was an English nationalist reaction against the formalities of Dutch William at Hampton Court and the grandiosity of the Sun King at Versailles, but the feeling ran deeper than that and beyond garden design. Burnet's 'stupendous primordial drama', his image of a globe broken asunder by divine anger and subterranean force, and his affirmation of great thoughts inspired by mountains, had worked on the imagination of succeeding generations. Edmund Burke in 1757 adumbrated a theory of the sublime and beautiful, the sublime inspiring a fear which fills the mind with great ideas and stirs the soul (such as a storm), the beautiful resting primarily on love and its associated emotions, refreshing the spirit in a different way. Already in the 1760s visitors were coming to the Highlands spurred by a wish to visit the 'sublime' scenery associated with the hysterical and tragic verse produced in James Macpherson's refashioning (faking, if you like) of the Ossianic tradition.[24] As Simon Schama has expressed it: 'Born from the oxymoron of agreeable horror, Romanticism was nursed on calamity.'[25] But it took the fastidious William Gilpin, born and bred in Cumberland, finally to blow the aesthetic improvement of nature out of the water by refining a theory of the picturesque.

It is easy not to take Gilpin seriously, to see him as a pernickety and self-important clergyman who set himself up as the arbiter of taste in the 1770s when enthusiasm for travel to mountainous regions came, in Professor Noye's phrase, 'suddenly and powerfully into fashion'. He was the original Dr Syntax of Rowlandson's caricature, and the butt of Romantic disgust for trying to contain nature in a picture frame. Yet, along with Edmund Burke's theories of the sublime and the beautiful, he had enormous immediate influence on his own generation, and this way of viewing nature through a lens of critical scrutiny lasted in principle, though not in detail, for the whole of the following century.

Let us follow him to the estate of the Duke of Atholl at the Hermitage of Dunkeld, just over the Highland line in Perthshire. He first takes the duke to task for his efforts to embellish the banks of the River Braan: 'his improvements are not suitable to the scene'. Nothing was needed but a path through the natural wood, but he has adorned it with an 'elaborate parterre' of knots of shrubs and flowers. One is about to turn back in disgust when we come 'unexpected upon an astonishing scene', the falls of Braan.

> This whole scene, and its accompaniments are not only grand; but pic-
> turesquely beautiful in the highest degree. The *composition* is perfect: but
> yet the parts are so intricate, so various, and so complicated, that I never
> found any piece of nature less obvious to imitation . . . This grand view
> which I scruple not to call the most interesting thing of the kind I ever
> saw, is exhibited through the windows of a summer house . . . too much
> adorned . . . Among its other ornaments, the panes of the windows are in
> part composed of red and green glass; which to those, who have never
> seen deceptions of this kind, give a new and surprising effect; turning the
> water into a cataract of fire, or a cascade of liquid vertigrease. But such
> decorations are tricks below the dignity of scenes like this . . .[26]

So we learn – and the duke learned, for he made adjustments – that it is nature unadorned that is beautiful, but not all nature, even unadorned, is picturesque: the composition has to be perfect.

There is also the question of notices. Gilpin found on his walk, and approved of, 'a gloomy cell, on the banks of the river', inscribed with a suitable verse from Ossian which 'joined its kindred ideas to that of the scene'. But he also found an inscription on a rock in the river bed, and climbing down to it in the expectation of 'an account of some life pre-served, or some natural curiosity found', read instead of a hole in the rock which had been 'drunk full of punch by gentlemen whose names are inscribed . . . one should have been sorry to have met the name of a friend recorded on such an occasion'.[27] There is a distinction, then, between information which explains nature and vandalism which disfigures it. We have it still.

The thrust of Gilpin's argument is to claim that the best of nature is above improvement. Wordsworth, that other and greater Cumbrian, was to take it all much further, maintaining that it was wrong to hold prissy

*The Great Auk: exterminated from Britain and the biosphere in the
nineteenth century. The last pair, killed in Iceland
in June 1844, were incubating an egg.*

theories as to how nature should appear – one should simply open a reflec-
tive mind to the impression of the wild, and let its mighty spirit work
upon the soul.[28] It was foolish and non-productive to compare one work of
nature with another: let them all speak. He was, of course, not consistent,
finding innumerable reasons why his beloved Lakes were superior to
anything in Scotland: in Loch Lomond, for instance 'the proportion of
diffused water is too great'.[29] But he liberated his fellow Romantics from
the mentality of the drawing room and the picture frame, and commanded
them to commune with nature, the wilder the better, so that ultimately no
spot of England or Scotland remained too desolate for admiration.

This sentiment reaches its apotheosis on Sir Walter Scott's voyage of

1814 with the Commissioners of Northern Lights along the northern and
western coasts of the Highlands, seeking the most sublime and terrific in
Scottish landscape. Scott awards the prize in his poem *Lord of the Isles* to
the extraordinarily remote and desolate Loch Coruisk on Skye, 'where the
proud Queen of Wilderness hath placed . . . her lonely throne':

> Such are the scenes, where savage grandeur wakes
> An awful thrill that softens into sighs;
> Such feelings rouse them by dim Rannoch's lakes.
> In dark Glencoe such gloomy raptures rise:
> Or, farther, where beneath the northern skies,
> Chides wild Loch Eribol his caverns hoar –
> But, be the minstrel judge, they yield the prize
> Of desert dignity to that dread shore,
> That sees grim Coolin rise, and hears Coriskin roar.[30]

So you can see what has happened. At virtually the same time as Walter
Scott provides his poetic shortlist of Areas of Outstanding Natural Beauty
– Skye, Rannoch, Glencoe, Northern Sutherland – the peat bog improvers
are talking about 'immense deserts . . . a blot upon the beauty and a derision to the agriculture of the British Isles'. Delight (raptures, in Scott's
stronger language) is one thing. Use (improvement) is quite another. Delight
is for poets, dreamers and women, use for practical men. In place of the
old unity constructing nature as static but simultaneously delightful and
useful, there is an alterable nature and a tension set up between use and
delight.

Peter Womack has argued in the Highland context that improvement
and romance, apparent opposites, are actually twins who ideologically
complement one another. Improvement is about maximising use 'for man',
and romance is about defining as delightful that which in economic terms
seems of least use, the archaic, the wild, the barren, the most 'natural'.
Because they are twins, he argues, they are not really in conflict. Romance
lives alongside improvement but does not oppose it, and 'nature is defined
as what we are not'.[31] Certainly the eighteenth century does divide man
and nature in a new way, and romance (or delight) was perhaps historically
timorous in opposing improvement (or use). But ultimately such opposition
did powerfully develop.

The struggle to defend the delightful was one into which Wordsworth
threw himself with enthusiasm, appearing now to some literary scholars as
a proto-green.[32] At first his interventions were on matters of taste, arguing
like Gilpin that nature must be respected. Thus to fulfil his garden designs
(Wordsworth once said he could have succeeded equally as landscape

Alfred Newton, FRS: friend of the great auk, Cambridge zoologist and father
of bird protection, painted by C. W. Furze in 1890.
Department of Zoology, University of Cambridge.

gardener, art critic and poet) William and Dorothy worked along with nature, moving wild plants as well as cottage plants into the gardens at Dove Cottage and Rydal Mount, searching the lakeside and fells for thyme, columbine, daisies, snowdrops, orchids, ferns and foxgloves in a way which would surely have appalled English Nature had it existed. Then he became involved in telling his neighbours what to do. In his *Guide to the Lakes*, the third section of which was entitled 'Changes, and Rules of Taste for Preventing their Bad Effects', he took *parvenue* settlers to task for felling native woods of holly, ash and oak and replacing them with Scots pine and larch (alien to the Lakes), embanking islands and building the wrong sorts of chimneys. He seems to have been among the very first to object to non-native species of tree, and he was a stern public defender of old woods and ancient individual trees which improvers wished to sweep aside.[33] Then, famously, and most futilely, towards the end of his life he opposed the railway's invasion of the Lake District, perhaps the first controversy in northern Britain between conservation and development. The railway directors won all the way to the bank, of course: how many divisions has the Poet?

Among the arguments advanced at the time in favour of the railway was that excursions would provide the working classes with much needed access to the countryside, to which Dorothy Wordsworth acidly commented that 'a greenfield with buttercups would answer all the purposes of the Lancashire operatives'.[34] There is nevertheless a direct line of agitation, voluntary action and public pressure, characteristic of northern Britain, between the Wordsworths' failed campaign against the railways of the 1840s and the National Parks and Access to the Countryside Act of 1949. The forces whose muscle was sufficient to bring the latter about were the very ones despised by the Wordsworths – working-class and middle-class ramblers and amateur naturalists, especially of the north of England, and the tourists and rural inhabitants (often retirees from the towns) who united in claiming a democratic right to enjoy the countryside for themselves.

This is not a story that I wish to tell in full here – parts of it are better left for the final chapter. But the line led through the ramblers' campaigns against landowners who, from the 1820s, were seeking to close rights of way, to James Bryce's unsuccessful series of Access to Mountains Bills, presented on eight occasions between 1884 and 1909, and through the Commons Preservation Society of 1865 to the foundation of the National Trust in 1894. It was no coincidence that it was threats and opportunities in the Lake District that led to the National Trust, or that the moving spirit of landscape conservation and access in the Lakes, Canon Rawnsley, was a devotee and disciple of Wordsworth. In 1907 a private Act of Parliament declared National Trust land 'inalienable', entailed, as it were, to the public for ever.

When National Parks legislation finally came in 1949 it did not apply to Scotland (except the nature conservation clauses in the act), and statutory protection of the beauty of the countryside through the designation of National Parks, Areas of Outstanding Natural Beauty and (later, in Scotland) National Scenic Areas has failed over the last fifty years to develop as strongly as statutory protection of wildlife through Sites of Special

Scientific Interest and, more recently, through the European designations of Special Protection Areas and Special Areas of Conservation. Both protection of beauty and protection of wildlife have in the twentieth century been mainly site-based, following the Enlightenment dichotomy that most of the world is for improvement but some of it can be romantically left for delight: most for man, a little for nature. Landscape protection, however, has been weaker than species protection, partly because, by definition, landscape is on a large scale and to protect it properly threatens to take large swathes of ground away from untrammelled improvement. It has also been weak because it has been seen to rest on subjective criteria: beauty is in the eye of the beholder.

Wildlife protection, on the other hand, has sought usually only to protect special creatures, until recently on rather small sites, and has claimed objectivity because it claims to rest on science. While it has often aroused bitter objections from developers, it has so far been the more effective because the rule of the bureaucrat guided by the scientific expert has been highly prized in government for most of the twentieth century. But it has been recently and persuasively argued that this may not last. If government becomes more open and democratic, particularly at the local level, it will become obvious that more of the public care about tranquillity and fresh air than about obscure species in the mud. Nature conservation, in Judith Garritt's words, has become a discourse between experts, of which locals do not feel part: 'the "non-experts" feel that their knowledge and perceptions are irrelevant, and that they are denied a role in deciding how the local environment should be used'.[35] Its relative, if narrow, strength is unquestionable in historic terms, but it could yet become its Achilles' heel, unless a wider popular excitement and understanding starts to grow about what scientists identify as special. Perhaps only birds have caught the popular imagination yet.

The history of harnessing science to the cause of nature conservation began in the Victorian age. Species preservation had no expression at all in the eighteenth century. To collect, to classify and to name was the extent of Linnean interest. Indeed, Linneaus himself had no concept that a species ever *could* become extinct. Species protection arises in the nineteenth century partly from the growth of feelings that Keith Thomas and Brian Harrison have tracked, that animals are sensate beings sharing the same earth, and the consequent banning of many cruel sports in the nineteenth century. An analogy was made between human experience and animal experience. When Charles Waterton returned from South America in 1813 and established his Yorkshire estate as Britain's first bird sanctuary, he said that it was because he had 'suffered himself and learned mercy'. Brigitte

Bardot, interestingly, explains her own interest in animal rights today in almost identical terms. A number of other landed individuals followed Waterton's lead. It was the Quakers and the Evangelicals that gave this approach particular support.[36] To that extent the origins of species' protection are non-scientific.

It also arises from collecting itself, and the appreciation that many organisms, insects and plants as well as animals were in danger of being driven from rarity to oblivion by the increasing host of amateur and professional naturalists in Victorian Britain: you may wish to kill the thing you love, but not to drive it to the point of extinction.

Important in formally involving scientists was, especially, Alfred Newton, first Professor of Zoology and Comparative Anatomy at Cambridge, who was deeply shaken by the twin role of human greed and scientific collecting in the extinction of the great auk, still relatively plentiful in the North Atlantic at the close of the Napoleonic Wars but exterminated when the last pair in Iceland was killed in 1844.[37] Newton had difficulty in accepting that the auk really could be extinct, went to search for it himself in Iceland in 1858, and was still writing in 1861 that if it were relocated it must immediately be protected by international cooperation against further depredations by collectors. Perhaps there could be a captive breeding programme in London Zoo. Given a second chance, modern science must 'transmit to posterity a less perishable inheritance' than 'mere possession of a few skins or eggs'. He foreshadowed the consensus of modern scientists to whom, in James Lovelock's words, 'the extinction of a species is as objectionable ethically as the burning of a cathedral or the destruction of some great painting'.[38]

Newton in 1868 addressed the British Association, warning that the continued slaughter of certain birds in the nesting season could lead to their extinction, and drawing particular attention to Flamborough Head. The problem here had attracted concern from local naturalists since the 1820s, as daily steamers took parties to shoot indiscriminately at the seabirds as they rose from the nesting ledges. It was changed from sport to business in the mid-1860s, when a change of fashion provided a market for gulls' wings on women's hats. Many thousands of seabirds were killed, according to the *Manchester Guardian*, 107,250 in four months, with up to 11,000 shot by eight people in a single week, with the wingless kittiwakes flung back into the sea to die and the young left on the ledges to starve. 'Fair and innocent as the snowy plumes may appear on a lady's hat', Newton told the British Association 'I must tell the wearer the truth – she bears the murderer's brand on her forehead.'[39]

Newton worked closely with the Revd F. O. Morris, who himself is a figure over whom it is worth pausing as he represents another tradition to

*James Bryce, OM: friend of the rambler, Liberal statesman
and father of the right to roam on the hills.
Scottish National Portrait Library.*

which modern conservation is profoundly indebted, that of the impassioned amateur observer. He was a considerable clergyman-naturalist in the Gilbert White tradition of natural theology. His works on birds and insects had a wide Victorian readership, combining good natural history with appropriate moral comment. He is fondly remembered now for commending as an example to his public the little brown dunnock, a bird of modest dress and self-effacing manners but now known to favour *ménage à trois* in its sex life.

Morris held that the glory of the Lord shone forth in the works of His creation, but he did not question that man had been given mastery of nature. Therefore, in the face of improvement for some general good of humanity he conceded that nature must step aside. The drainage of the fens was, for him, a triumph of hydraulic science, 'a mighty and very valuable victory, and the land that was once productive only of fever and ague now scarce yields to any in the weight of its golden harvest'. Only the entomologist, he went on, had cause for regret, 'and he, loyal and patriotic subject as he is, must not repine at even the disappearance of the Large Copper butterfly in the face of such vast and magnificent advantages.'[40] Exactly this kind of diffidence, where the lover of delight felt disempowered by the hegemony of use, reached well into the twentieth century, as when in 1935 Campbell Nairne, writing about a new hydro scheme, felt 'numb with horror' at his first sight of the 'black pipes that lance the loins of Schiehallion' but: 'it had to be, and one accepts the spoilation of the country . . . as the toll levied for progress'.[41]

But if Morris was prepared to take it when pumping engines wiped out the works of God, he certainly was not when naked callousness and greed appeared on the cliffs of Scarborough armed with a gun and similar intentions. Newton, Morris and the local clergy, under the patronage of the Archbishop of York, formed a Yorkshire Association for the Protection of Sea Birds, possibly the first wildlife conservation body in the world.[42] The upshot was also the first wild bird protection legislation on the British statute book, the Sea Birds Protection Act of 1869, which imposed a close season from 1 April to 1 August for thirty-three kinds of seabirds.[43]

Students of British social legislation in the nineteenth century will not be surprised to hear that the act and its immediate successors were largely ineffective, vitiated by exemption clauses which allowed egg-collecting and local depredations to continue in the name of tradition. But Newton and Morris had forged an alliance between scientists and the new moral view of nature entailing responsibility for other species. The next Act, of 1872, contained a list of eighty-two birds recommended for protection by a committee of the British Association at the prompting of Newton: the long process of scientists instigating and validating schedules of protected species had begun, though some of those mentioned in the 1872 Act, like the plover's page and the summer snipe, might puzzle modern taxonomists.[44]

The rise of nature conservation as a powerful force in the twentieth century owes everything to the twinning of science and voluntary effort which Newton and Morris began and exemplified. It has continued in particular to involve the protection of birds, associated with the rise and rise of the RSPB, from its foundation in 1889 by ladies sworn to oppose wearing plumage (ostrich and game feathers excepted), to its transition to a body involved in protecting wild birds in every way (marked by a grant of a Royal Charter in 1904), to its maturation in the second half of the twentieth century into the most powerful nature conservation body in Europe, with a million members in 1997. But it equally involved two bodies of a very different kind, both founded in 1912, the British Ecological Society, with Professor Arthur Tansley as its first President, and the Society for the Promotion of Nature Reserves, with N. C. Rothschild as its moving spirit. The close alliance of scientists and amateurs now involved a search not only for species to protect but, more importantly, for habitats and localities to defend.

It led eventually, largely due to the BES, to the establishment in 1949 of a government biological service with responsibilities in Scotland as well as in England, the Nature Conservancy. Prime among its responsibilities was the establishment of a network of National Nature Reserves and Sites of Special Scientific Interest. John Sheail has said of it that it transferred the question of nature conservation from the planning to the science sector of government,[45] but it became too much of a hot potato to remain in the lofty province of science alone. Particularly after the 1981 Wildlife and Countryside Act offered the network more real protection if the price of compensation could be met, the successor bodies to the Nature Conservancy have often found themselves embroiled in bitter planning controversies with land managers and developers.

It is ironic that research scientists for many years have shown little interest in the vast majority of Sites of Special Scientific Interest: as far back as the 1960s, in a famous dispute over the flooding of Widdybank Fell to construct Cow Green Reservoir in Teesdale, the case was lost by the Nature Conservancy because the arctic-alpine flora, wonderful and irreplaceable though it was, could not be shown to be of substantial significance to the advancement of science.[46] It is mainly the amateur naturalists and the informed public, the descendants of Morris rather than of Newton, to whom they are interesting and to whom SSSIs give delight. But the choice of what to designate remains underpinned by professional scientific judgement, which gives nature conservation a particular authority in the eyes of government and (so far) the populace. The advantage for species and habitat protection, compared to landscape protection, is that in this sphere nature can be construed as resting on objective science, whereas landscape protection is an area construed as resting on the wavery concept of beauty, where one man's meat is another man's poison. Consequently, species and habitat protection in modern legislation is a great deal stronger.

Amateur botanising party on Ben Lawers, Perthshire, 1925.
St Andrews University Library, Adam Collection.

Use and delight in the twentieth century have nevertheless come to be in endless conflict. Let me take grey seals as an example that draws together some of the threads. Before the nineteenth century there was no concept of species protection, and no sense of delight in seals as far as historical or literary sources allow us to judge, unless you consider the folk tales and superstitions surrounding 'selchies' in the Hebrides and northern isles themselves to be a source of delight. There was plenty of use, of their skins, flesh and oil. Thus the inhabitants of Arbroath in Angus in the seventeenth century hunted the seals, 'the old ones are of a hudge bignes nigh to ane ordinare ox but longer'. The hunters went into the breeding caves 'with boates and with lighted candles' to kill young and old, 'whereof they make very good oyll'. Their contemporaries in North Uist visited the breeding islands of Haskeir in calm weather, and bring 'bigg trees and stafs in their hands with them as weapons to kill the selchis . . . and so the men and the selchis doe fight stronglie and there will be innumerable selchis slaine wherwith they loaden their boatts, which causes manie of them oftymes perish and droune in respect they loaden their boats with so manie selchis'. And in Orkney the inhabitants of North Ronaldsay similarly attacked Selchskerry, armed with hazel sticks, and 'the monsters, eyeing them with dread and gnashing their teeth with rage strive to get out of the way with wide open mouth, then they attack with all their strength'. If the first escaped without injury 'all the others attack with their teeth', but if they fail 'all the other take to flight and are easily captured, and I have seen sixty taken at one time'.[47]

The grey seal was a formidable adversary to a man armed with a stick, and in due course man improved the odds in his favour. Charles St John has a vivid and unpleasant account of seal hunting in the north of Scotland in the 1840s when the panic-stricken animals were herded over iron spikes driven into the rocks, and those that escaped disembowelment shot with guns.[48] Increasingly grey seals sought the remote and uninhabited islands like North Rona, but sealers followed them there. In the new mood of compassion towards animals, some individual landowners, like the proprietor of Haskeir in 1858, tried, without a great deal of success, to protect colonies on their own property. By the start of the twentieth century the population had fallen steeply, and there was considered a real risk of their extinction in the near future. Sealskins were particularly fancied as driving jackets for the flashy new motorists.

E. F. Warburg and Grant Roger from Nature Conservancy examining a seedling of the endemic whitebeam, Sorbius arranensis, *Glen Diomhan, Arran, 1958. Thomas Huxley.*

These were the circumstances under which the Grey Seals (Protection) Act of 1914 came into force, the first time a large mammal, let alone one with economic importance both as a resource and as a predator of commercial fish, had been given statutory protection. It was forbidden to kill them in their breeding season for a trial period of five years; in the event the Act remained in force, on scientific advice, until 1932 when it was strengthened and made permanent. A Scottish population allegedly only 500–1,000 animals in 1914 had reached 4,300 by the late 1920s: by 1963 it was 30,000, by 1978, 60,000, and in 1996 was estimated at 111,600 and increasing at the rate of over 6 per cent per annum.[49]

For the seal at least this is an extraordinary success story, especially as the Scottish population represents three-quarters of the world population of what is still one of the world's rarer mammals. It has been achieved by a combination of science and popular feeling. A special body of scientific advisors, the Sea Mammal Research Unit, has been funded by government since the late 1970s to advise it on matters relating to seal conservation. And the public greets any suggestion to reintroduce a cull of seals (as was attempted last in Orkney in 1978) with such loud hostility that politicians back away. Seals have developed an iconic appeal as a source of delight.

The steps in this transformation from quarry to icon are obscure. The first talking seal in literature (as opposed to folk tale) was the 'Walrus and the Carpenter'. In 1863, Charles Kingsleys' *Water Babies* gave a splendid account of an aquatic ecosystem populated with talking animals. Trips to see common seals on their breeding rocks were increasing as seaside holiday attractions before the Second World War, and Frank Fraser Darling in his study on the grey seals on North Rona in 1943 noted 'there is no creature born, even among the great apes, which more resembles a human baby in its ways and cries than a baby seal'.[50] Kempster, writing *Our Rivers* in 1948, already feared that this anthropomorphism would come to prevent what he regarded as rational pest control. In the 1970s, Greenpeace took up the photogenic cause of helpless baby harp seals bludgeoned to death by Norwegian sealers on the Canadian pack ice, and with a little help from Brigitte Bardot made startling international television. Sealife and sea-rescue centres in Britain increasingly made children and adults familiar at close range with injured seals awaiting return to the sea.

So to kill seals simply became taboo to most of the British public in the 1970s: they were judged too delightful for use or destruction of them to be justified. But this is not a universal human, or even Western, phenomenon of the late twentieth century. In Norway seals are not only still shot when they appear on the coasts or at the mouths of rivers, but government subsidies are still paid to sealers in the north of the country to go to the pack ice and club the young. Their skins are now unsaleable, but they are warehoused at public expense. The grounds for this – that sealing is traditional – seem at least as irrational as the grounds for making seals taboo. But culture is no less powerful for being irrational, and the public has every right to insist that beautiful and sensate animals that look sometimes a little like themselves should be shown veneration and respect.

Unfortunately, grey seals remain a source of controversy in northern Britain, since not everyone is convinced of their new status. Fishermen allege that the inflated population is causing serious damage to salmon, cod and other stocks of fish. They call for a renewed cull, but their demand to be allowed to sell seal penises to the Chinese as aphrodisiacs or their flesh as pet food for the supermarkets seems to show a rash disregard for wider British sensibilities.[51] Nevertheless, the Scottish National Party, strong in the fishing constituencies, does support a new cull, deploring 'environmental Luddites' who ignore the consequences of the recent increase in seal numbers and declaring that 'the days when jobs are sacrificed on the alter of political correctness are gone . . . seals are attacking salmon . . . salmon are a national, natural asset'.[52] This perspective conveniently omits the role in the recent precipitous decline in wild salmon and sea trout of job-creating fish farms by the mouths of the Scottish rivers, and of deep-sea fishing itself. It is simpler to blame the seal.

The Labour administration, however, still after twenty years recalling the retreat of the redoubtable Secretary of State, Bruce Millan, in the face of public anger after the last culls on Orkney, prefers the status quo. The minister, Lord Sewel, with academic calm, emphasised the need for better scientific evidence of substantial damage, pointing out that it is still legally permissible to shoot a seal attacking salmon, nets or tackle. He did not say how hard it is to hit one.[53]

The scientists, for their part, remain convinced that the fishermen's claims of seal damage are exaggerated but worry about the sustainable nature of population growth at this rate. The truth is that the one predator that the seal has had since time immemorial, the human species, has voluntarily removed itself from the scene. To have done such a thing is a consequence of our culture, in this century, constructing nature in a particular way. It would be a misrepresentation to say that science approves or disapproves the outcome, but historically it came about because scientists endorsed the need to save the seal, and popular culture then adopted the seal as a totem animal.[54]

So, delight in seals is very real and highly politicised. But, as so often, those who show delight are not those who are affected by the policies to control or prevent use. Nature is contested ground because the relationships

and the confrontations between use and delight are real, because one side's
totem is the other side's pest, and not all the arguments are on one side. We
would not be sitting here unless we had exploited nature along the lines
proposed by Bacon and the Enlightenment: it is the foundation of economic growth, and the root of the ways in which we usually measure GNP.
Yet, as economists are increasingly urging us to realise, there are more
measures of welfare than GNP alone, and access to a decent environment,
delight in nature, is for many as real a component of the standard of living
as access to money. For a long time this has been so. What are the costs
of growth and where should its limits be? The question which Ruskin
repeatedly asked in several forms remains completely relevant: 'If the
whole of England were turned into a mine, would it be richer or poorer?'[55]

We shall see in the remainder of these chapters how society has tried to
find a balance between use and delight that tries to maximise benefit to us
as a species, and how nature itself bears the impress of this endeavour, and
may yet cause us to revisit our judgements, and our misjudgements.

CHAPTER 2

WOODS OF IMAGINATION AND REALITY

Let us begin with the Great Wood of Caledon. It is, in every sense of the word, a myth. Much of the talk of forests in the modern world is about catastrophic destruction, the end of the indigenous tropical hardwood forests as they go up in smoke in Brazil or Indonesia, or the doom of the old-growth coniferous forests in North America and Siberia swept into the pulp mills of Japan and the West. A powerful tale is told in Scotland, often on television or in the press, of similar catastrophes. Here is an extract from the recent book by Hugh Miles and Brian Jackman which accompanied the former's prize-winning film:

> The Great Wood of Caledon is our oldest British woodland, a primeval northern forest which had already been standing for at least 2000 years when Stonhenge was raised . . . there was hardly a glen that was not roofed with trees, the high hills rising like islands from among the blue-green canopy. The wolf and the lynx roamed its trackless deeps. Bears and wild boars snuffled among its roots . . . When the Romans came to Britain it became a refuge for the Pictish tribes who waged guerrilla war on the imperial legions . . . Then came the Vikings, and the war on the Great Wood began . . . they torched the forests, and felled the tall trees to fashion masts for their longships . . . For the Highland Scots the trees were fuel, and huge gaps appeared in the canopy as feuding clans burned the woods of their enemies. Yet still the Great Wood stretched for miles, a sanctuary for wolves and renegades alike until the English arrived to smoke them out . . . the crushing of Bonnie Prince Charlie's Highland rebels signalled the end of the glory of Caledon. Down came the mighty trees, felled by the impoverished clan chiefs, who forced to pay off their hated Hanoverian landlords, sold their timber to English ironmasters.[1]

This is history going off the rails, starting more or less true and crashing in hopeless error. Yes, there was a time at about 3000 BC when Scotland, like Ireland, was covered with forest, barring only the mountain tops, the marshes and the standing water, though it was not all (or even predominantly) a forest of the blue-green canopied Caledonian pine. It was variously of pine, birch, oak, elm and hazel, in differing composition in different places, with most of the conifer in the north and east of the Highlands. The many surviving semi-natural remnants of this forest in Scotland are no older than those of various broadleaf woodland in other parts of the British Isles.[2]

In a European context there was little that was remarkable about this forest: much of the northern hemisphere was similarly covered, though in the boreal forests of Scandinavia and Siberia there was a sharp distinction between the spruce on the wetter lands and pine on dry soils, while in Scotland after the last Ice Age there was no spruce and pine struggled to occupy the wetter niche. Yes, wolves and boars, and probably bears and lynx did roam its depths.[3] By the time the Romans arrived, the forest had already been in retreat for thousands of years, falling back under the natural onslaught of climatic change as a shift of polar winds to the south in the Bronze Age brought about a critical switch to greater oceanicity, with heavy rain, strong winds and incessant peat formation. The absence of spruce meant that there was no conifer well adapted for a wet environment.

The blackened stumps in the bogs, which look so fresh when dug out in the peat workings or encountered by the hill walker, are usually dated by dendrochronology or carbon analysis to about 4000 years ago – relicts of this great climatic change. On fertile ground, especially in the Lowlands, much of the surviving broadleaf wood was felled and cleared by Celtic tribes in the first millennium before Christ, if not before.[4] Even on the hillsides, the extent and character of the wood must have been profoundly altered by several millennia of tribal grazing animals. According to David Breeze, Historic Scotland's Chief Inspector of Sites and Monuments, the only time that we can be sure that the Romans actually encountered a wood was when the enemy retreated into one after the Battle of Mons Graupius in AD 84, and the battle itself took place in the open, with the Caledonians using chariots.[5] The Vikings and the medieval clans inflicted no known damage from setting fire to the pine woods: indeed, had they done so it would have provided a good seedbed for regeneration. They may well have had an impact from their grazing animals and from repeatedly burning heather moorland to maintain a rough pasture, but that is another matter. Where peasant use of this kind was heavy, it would have converted wood to open wood pasture, impoverishing its species base, confining small trees like holly and roses to the stream sides, and (above the tree line proper) causing montane willow and birch to decline. Ultimately it would have made bare moor.

The impact of the English after the '45, dare I say it, was mainly benign. Such speculative raids on the woods as they and the Irish indulged in were generally unsuccessful, virtually all in the previous half-century and entirely at the instigation or with the active cooperation of the chiefs. After the rising not only did the chiefs *not* have hated Hanoverian landlords (a few were forfeited, most left in peaceable possession) but the much maligned ironmasters purchased oak coppice, not mighty pine, and stimulated an improvement in its management without parallel in Highland history.[6]

I dwell on Miles and Jackman not from an arsonist's wish to burn straw men, but because statements of this kind have a very wide currency both in Scotland and England and have formed a construction of environmental history that has no small psychic and political energy. They have done so at least partly because they have received the *imprimatur* of scientists. Thus the very influential ecologist, Frank Fraser Darling, in 1947, told the story very much along the same lines as Miles and Jackman were to do, again implying that the main decline is only a few hundred years old and concluding with a moral about the Scottish woods of his own:

> Man does not seem to extirpate a feature of his environment as long as that natural resource is concerned only with man's everyday life: but as soon as he looks upon it as having some value for export – that he can live by selling it to some distant populations – there is real danger.[7]

So the Great Wood was a lesson about capitalism as well. Darling came to be regarded in due course as a popular ecological guru, a native Aldo Leopold. For example, in the 1980s, the group Reforesting Scotland had influence not only on the Green Party, but also on the much wider forestry movement towards planting native broadleaf and Caledonian pine woodland in place of Sitka and lodgepole. Their chairman was only restating Darling's opinions when he spoke of the extirpation of the Great Wood of Caledon as leaving the Highlands 'almost totally deforested and in the final phase of vegetation and soil degeneration'.[8] Nor was this the whole story. As early as 1961, Mark Louden Anderson, Professor of Forestry at Edinburgh University, told a symposium that:

> The financing of afforestation or reafforestation is not easy to justify by any orthodox argument. It can, however, be amply justified when it is regarded as a task of reparation or restitution.[9]

The industry as well as conservationists needed the Great Wood. Perhaps Sitka spruce could even stand in for the original. In 1973, a forester at a natural resources symposium cited a 'personal communication' from Fraser Darling describing Sitka as a 'godsend' to Scottish hill ground and as the means ultimately to bring about a new 'forest biome after a long period of soil degradation'.[10]

This moment passed, however, and critics of the forestry industry

Imagination: frontispiece of the Sobieski Stuarts'
Lays of the Deer Forest, *vol. 2.*

Reality: old Scots pine among birch, Glen Affric, Inverness-shire, 1930.
St Andrews University Library, Adam Collection.

decided that the Caledonian forest could never be truly replicated by
serried ranks of alien trees. There had to be an attempt at restoration. That
the Millennium Forest Trust for Scotland, composed of many interests
coming together, obtained in the 1990s several million pounds of lottery
and other public money for the wholly admirable project of planting and
restoring native woodland was due in no small way to the grip on the
public imagination of the Wood of Caledon.

So how did the myth arise? Of course it does originate with the Romans.

Pliny in AD 43 remarked that the Roman army had not penetrated beyond the Caledonian forest, but does not say where that was: it seems to have been like Ultima Thule, not a real place but 'somewhere up there'. Ptolemy in the next century records that the Caledonian forest was 'beyond' (or possibly 'west of') the Caledonii, a tribe whom they had now just encountered in an area north of the Forth: so it was still somewhere beyond the known horizon. Tacitus, Dio and Heroditan in describing the campaigns of Agricola and Severus say that their troops, in order to get to grips with the enemy, had to wade through marshes and fight through woods. Unfortunately, this is all in general terms and turns out, when examined in context, to be part of the conventions of Roman literary style, tenuously related to reality. Dio, for example, also describes the Caledonians and their fellow tribesmen, the Maeatae, as inhabiting 'wild and waterless mountains', as pastoralists with no other knowledge of agriculture, and as able to live for days immersed up to their necks in water. He also says that Severus not only cut down the forests but drained the swamps and levelled the hills – all claims hard to check by his readership in Rome.[11]

Archaeology, however, demonstrates that in reality the Romans occupied in Northumbria and the Scottish Lowlands a relatively unwooded country, exploiting native tribes who possessed cattle, sheep and extensive arable cultivation. They made use of timber in their constructions, but also of stone and particularly of turf. Hadrian's Wall and the Antonine Wall, let alone the line of signal stations at Gask in Perthshire, would make no sense unless they looked out over open country. The Iron Age tribes may have been as numerous as the population of medieval Scotland and had had a long time to change their world.[12]

About Highland wood cover in Roman times we know less. It was probably more plentiful than today. It was certainly discontinuous, and almost everywhere would have been subjected to modification by climatic change and early peoples. At Balnaguard in Perthshire, near the confluence of the Tummel and the Tay, for example, there are signs of interference with the natural vegetation as early as the seventh millennium BC, and traces of cultivation before the late third millennium. In remote Glen Affric in Inverness-shire, long-term pastoral activity with some cultivation probably began during the early Bronze Age and occurred through the Iron Age, though in West Affric the tree cover had already gone through natural causes, above all increased rainfall and peat spread, about 4,000 years ago.[13] David Breeze, possibly overstating the case, goes so far as to say that the Highland woods had 'been cleared 2000 years before the Romans arrived, in the Neolithic and Bronze Ages, leaving only isolated pockets of the original pine and birch woods'.[14]

But it was not directly from Roman primary sources that Fraser Darling and others drew their main inspiration, but from the late medieval Scottish chronicler, Hector Boece, the first principal of Aberdeen University. Like the Roman authors, posterity read him selectively. It is worth recalling that he also believed that barnacle geese hatched from seashells. In 1527, he described the *Caledonia Silva* of the Romans as a great wood that once had stretched north from Stirling, covering Menteith, Stratherne, Atholl and Lochaber, full of white bulls with 'crisp and curland mane, like feirs lionis'.[15] He set it firmly in the past but gives no hint that he thought it had been dominated by pine: in his day, and earlier, the dominant tree at least in the south of that area was oak. The English topographer Camden embroidered Boece for an Elizabethan audience, and the 'Caledonian wood' became an 'immense tract of ground impervious for thick trees'.

Seventeenth- and eighteenth-century commentators either do not mention it at all, or give it a reference in passing. Thus to Sir Robert Sibbald in 1684 it had long disappeared apart from obscure vestiges.[16] To a French cartographer in 1708 *Caledonia Silva* still existed but was quite small; he named a wood in Glen Orchy by this title. George Chalmers, in 1807, was the first to go back to Roman sources in detail and repeated the improbable story of Severus ordering his soldiers to fell all the forests in Scotland, commenting sensibly that this labour would have exhausted them and contributed to their subsequent defeats at the hands of the Picts.[17]

Of course, by then plenty of people in the south knew about extensive if scattered woods of pine and oak in the Highlands, and some considered them an important actual or potential economic resource. But only Lachlan Shaw in his *History of the Province of Moray* conceptualised them in one sentence as perhaps 'the remains of the Sylva Caledonia' and even he did not invest them with any historic importance or particular national symbolism. The observant and well-informed Thomas Pennant made no mention of a former Great Wood, though he listed many of the surviving native pinewoods and was an appreciative visitor to the forests of Deeside and Black Wood of Rannoch.[18]

Nor, perhaps surprisingly, did the Romantics make much of the Great Wood of Caledon, though they were aware of early references and certainly admired the picturesque effect of the native pines. Instructed tourists like J. E. Bowman, a literary banker from Wrexham who did the Highland circuit in July and August 1825, knew the difference between the 'large detached and scattered groups of trees' of Strathspey and the plantations of the south and recognised that the former 'were the Natural Pine Forests'. When he came to Atholl he called it 'part of the Caledonia Sylva', long 'despoiled of its honours, and with them its wild bulls with thick manes, its witches and wicked women, and other hideous horrors spoken of by Camden'.[19] That was about as far as anyone went before the 1840s.

In its modern form, the Caledonian Forest is a product of German Romanticism, mediated through the excitable and fantasy-filled minds of the Sobieski Stuarts. These two brothers of Anglo-German descent falsely claimed to be the legitimate grandchildren of Prince Charles Edward, though, reassuringly, they did not wish to press a claim to Queen Victoria's throne. They did, however, set themselves up to an admiring public as experts in the Gaelic past, claiming to have a sixteenth-century volume describing clan tartans which, rather like Macpherson and the Ossianic manuscripts, they never could produce in reality. Among their publications was, in 1848, *Lays of the Deer Forest*, a compilation of poetry, passionate and breathless hunting tales, natural history, folklore and popular history, often only tenuously anchored in verifiable fact. They spoke of a great forest, much of it surviving until recent times, that had once occupied the province of Moray.[20] It had been filled with remarkable beasts, wolves, bears, boars, beavers, wild cattle and giant red deer, but even this was only 'a skirt' of *Caledonia Silva*: 'the great primeval cloud which covered the hills and plains of Scotland before they were cleared.'[21]

The Sobieski Stuarts told a wild tale of Jenny Macintosh, lost in Rothiemurchus while gathering pine cones and found dead beneath a great pine three years later. It was very like something out of Grimm, emphasising the disorienting and fearful character of the wood: 'some shreds of her plaid . . . were in the raven's nest on Craig-dhubh, and a lock of her grey hair . . . was under the young eagles in the eyry of Loch-an-Eilean'. The unwary would never have suspected that Rothiemurchus was among the longest exploited and most populous of Scottish forests. They also larded their book with impressive references to aristocratic hunting in Hungary, Bohemia and Austria.

Their style reflects the mind-set of the German forest romantics, men like Wilhelm Pfeil, whose 'deep emotional adherence to forests' was closely connected with his pleasure in hunting, like Ernst Moritz Arndt, to whom the axe which fells the tree threatens to turn into the axe which fells the whole people, and like Wilhelm Heinrich Riehl, for whom the wild wood (and all German forest to him was wild wood) was the very foundation of German freedom. As Simon Schama has aptly remarked, 'religion and patriotism, antiquity and future, all come together in the Teutonic romance of the woods'.[22]

Further, more academically respectable, impetus to the developing tale was given by W. F. Skene's pioneering study of Highland history, *Celtic Scotland* (1871), where he invites the reader to compare Severus's campaign

The royal hunting park of Falkland Wood, with Falkland Palace, Fife,
painted by Alexander Kierincz, c. 1636.
The enclosure paling of 'steak and rice' can just be made out.
Scottish National Portrait Gallery.

against the Caledonian tribes to that of Sir Harry Smith against the Kaffirs
of Cape Colony in 1852, quoting the Duke of Wellington on the need to
drive roads through the jungle and the effectiveness of guerrilla warfare
against the British army in the bush. Much of Roman Scotland, he says
without giving evidence, must 'have presented the appearance of a jungle
or bush of oak, birch or hazel'.[23]

Skene's invitation was then taken up by David Nairne, in 'Notes on
Highland Woods' of 1892. Most of his essay is marked by excellent scholar-
ship that represents the start of serious research into Scottish woodland
history, but it begins with a completely imaginative (and Teutonic) flight
into the Great Wood of Caledon, which he describes as covering the entire
Highlands:

First comes Scotland in its primeval grandeur of mountain, forest, and
flood, the war cry of the sturdy aborigines finding an echo in the woods
wherever the tribal battle was waged, or the shout of the barbarian sportsmen

as they merrily with bow, sling and lance, pursue the crusade against the
wolves, and the bears and the reindeer. Here is Scottish freedom in embryo.

Then he goes on to describe the Roman invasion, the natives gather in
the 'rude panoply of war', blood flows freely, 'but steadily the Roman
legions cut their way through the pathless tracts of Strathspey...
Victorious, but at what a cost!' The native guerrillas withdraw 'happy only
in the thought that 50,000 men of the invading hosts have fallen as the trees
they felled', and so on.[24]

So Miles and Jackman's book was almost written for them by the
Sobieski Stuarts in 1848 and David Nairne in 1892. The myth needed only
a few misjudgements about the Middle Ages and the English to achieve
modern form as an example of the terrible abuse of the wild wood and the
hard lot of the Scots, for both of which it behoves the modern world to
make amends.

So let us try to take a more measured view of what really happened to
the woods of northern Britain, at least over the last four or five centuries. I
need to make the gist of my reinterpretation clear at this point. I agree that
there was once a vast wood covering Scotland – but that was 5,000 years
ago, not in Roman times or since. I agree that there was substantially more
ancient or semi-natural wood at the end of the Middle Ages than there is
today, but not that Scotland, even in the Highlands, was then a notably
wooded country by European standards. I agree that this remaining wood
has also declined severely and continuously since 1500, but not that the
decline has been concentrated since industrialisation. I do not agree that
those primarily responsible were outsiders, and overall I attribute much of
the decline to natural forces and to the pressures of indigenous peoples
and their flocks. One aspect of the story which is left until later is that of
montane scrub, dwarf birch, willow and so forth, which may indeed have
been much more plentiful in the eighteenth century than today.

First, how much woodland was there? The first reasonably comprehensive evidence comes from Roy's Military Survey of the 1750s and suggests
that about 4 per cent of the land surface of Scotland was forested, very
little of that being plantation. The next estimate, of 1815, suggests about
3 per cent wood cover, and two further estimates late in the nineteenth
century suggest between 3 per cent and 4 per cent. By 1914 the figure was
still under 5 per cent, and today is about 20 per cent though of that under
2 per cent is semi-natural woodland. Northern England was much the same.[25]
Estimates of woodland cover are always open to question, especially the
earlier ones, but even if they were out by a margin of 50 per cent it would
not make much difference.

The trends are obvious. Firstly, the percentage of land covered by wood in Scotland was low in European terms from the mid-eighteenth to the mid-twentieth century. Secondly, over the last 250 years one-half to two-thirds of the semi-natural wood has been replaced by plantation. Thirdly, and most importantly, if, at the woodland maximum 5,000 years ago, 80 per cent or more of the land had been covered by wood, then 76 per cent had been deforested before the Industrial Revolution. That alone makes mince-meat of the argument that greedy modern capitalism was to blame.

How much disappeared between the end of the Middle Ages and 1750? This is a much more uncertain question, but a cartographic snapshot of Scotland is provided by Timothy Pont's sketch maps of much of the country in the 1580s and 1590s, and a topographic one by the investigations of the seventeenth century carried out by Sir Robert Gordon and Sir Robert Sibbald.[26] The impression from these is that there was more woodland in the late sixteenth and early seventeenth centuries than there was by 1750, but that Scotland was already relatively sparsely wooded, with more in the Highlands than in the Lowlands, but even in the Highlands with thin and discontinuous cover in most places. I would guess that the total woodland cover could have been 10–15 per cent of the land surface in 1500. All the large mammals except the wolf had vanished: the auroch and the reindeer in prehistory, the lynx, the elk and the bear before the onset of the Middle Ages, the wild boar before their end, the beaver by 1550.[27] The wolf, still plentiful in the north in Pont's day, was virtually extinct one hundred years later. All this points to the conclusion that the rate of erosion of woodland was as great or greater in the centuries leading up to the Industrial Revolution as it was later.

Whatever the cause, it was certainly not due to blind neglect, particularly not in the southern half of our area. In the Middle Ages there had developed in the Scottish Borders and north of England, without any conscious plan, a division of land use which left most of the uplands to commercial sheep grazing, tilled the most suitable of the low ground and preserved only what wood was necessary to support agriculture, turnery, leather-working and metal-working. The effect had been steadily to reduce the woodland cover of Cumbria, Northumberland, Yorkshire and the Scottish Borders, but not to the point of extinction or even economic inconvenience.

The Lake District provides a good example. 'In Cumbria', explains Dr Winchester, 'the history of woodland over the medieval centuries appears to be one of destruction until the demand for charcoal led to more active management in the sixteenth century.' It was ultimately the greater use of the woods for iron-smelting, lead-smelting, tanning, turnery and making potash for the textile trades that came to the rescue of the woods: 'The perception of woodland was changing: no longer was it a survival of the untamed landscape to be exploited at will, but rather a dwindling resource, valuable for a range of uses and requiring careful protection.'[28] Relative scarcity led to this protection, enclosure and more systematic coppicing, even the planting of new woods on land used otherwise for rough grazing.

The increase in attention meant that more wood survived, but not that it survived in an ecologically pristine state. About 35 per cent of the area around Coniston Water remains under semi-natural oakwood (the average for Cumbria as a whole is about 5 per cent). This is entirely due to the maintenance of coppice in the early modern period, used to make charcoal for the local iron and gunpowder industries, for tanbark, for structural timber and for 'swill-making' – the art of weaving split oak into a basket. Most of the woodland has also been subject to unrestricted sheep grazing for a century or more following the collapse of these industries. The removal of wood products over centuries has, in Susan Barker's view, 'probably been the major contributing factor to the nutrient impoverishment of wide areas of the Coniston woodland'. The most biodiverse areas today are the inaccessible gorges along the streamsides, which suffered less from coppicing and grazing: only here do trees like small-leaved lime and herbs like wood fescue and tutsan survive. The remaining woods between the streams have been depleted by centuries of intensive use, but at least they are still woods.[29]

This story of management and modification was true everywhere. In north-east Yorkshire, there was variety of treatment of the woods on the very large Helmsley estate of Lord Villiers in the seventeenth century: some woods were valued for coppice, some for mature standard oak, some for ash, some for a medley of different things depending on their locality and economic functions.[30] Up in Swaledale on the Pennine edge, meanwhile, extensive woodlands in retreat by the end of the thirteenth century had become, firstly, late medieval wood pasture, partly converted to deer parks, and then in the sixteenth and seventeenth centuries cow pastures of ever-decreasing tree density, but leaving pollards to stand among the fields and provide timber and herbiage.[31] In these uplands, away from an outside market and with plenty of trees for local use, there was less care to protect woodland from competing agricultural use.

The history of Scottish woodland of the seventeenth and early eighteenth centuries has been less closely studied than that of northern England, but at least in the Lowlands, the vocabulary of exploitation was very similar. Deeds speak about sales of wood in 'hags' – divisions which were felled sequentially by the buyer; they enjoin responsibility to enclose the hag at the end of the year with a fence of 'stake and rice' (essentially wattle) to keep animals from harming the regrowth. They have rules about coppicing – it is to be done close to the ground, the stool is not to have a dished surface where rain could collect and peeling bark below the cut of the axe is disallowed. Frequently, a certain number of trees were reserved from cutting to grow on as standards. As well as internal enclosures there was an external enclosure maintained by the landowner, often of earth topped by stake and rice, sometimes a stone dyke.[32]

As in England, however, there was great variety in what actually went on in the woods, and the degree to which coppicing in Scotland was on a systematic and predetermined rotational basis remains obscure. It clearly became more systematic as time passed. In the early eighteenth century oak woods on the Duke of Montrose's estate in Menteith were managed for cutting on a twenty-five year rotation, and by the end of the century many oak woods throughout Dunbartonshire and Stirlingshire were similarly husbanded, mainly for tanbark. But even in the seventeenth century there was no doubting the value and care bestowed on many of the woods. Edinburgh, for example, drew on a half circle of woods to the south and east, including Ormiston, Humbie, Pressmennan and Roslin, where the felling rights were bought by the capital's merchants and tanners according to carefully specified rules: as a group, these woods were probably the most valuable in Scotland at that time.

It is of course true that large areas of the east coast of Scotland, for example Buchan and Fife, were largely destitute of trees – a point noted by native topographers and visitors alike. This, however, was of little disadvantage. For fuel the inhabitants had abundant peat or, in Fife's case, coal and peat – anyway, over most of northern Britain wood was seldom essential for heating or cooking. For tanning, Fifers could import materials from elsewhere in Scotland. For building, they had access to the abundant pine of Ryfylke and Sunnhordland in Norway, only about three days sail from the Firth of Forth and involving much lower transport costs than trying to extract wood from the Highlands. Indeed, right down the eastern side of Scotland and England the easy access to Scandinavia and the Baltic obviated the need for home-grown building timber in any of the port cities.[33] As Adam Smith observed, in the whole New Town of Edinburgh you would not find a single stick of Scottish timber.[34]

But, just as in Cumbria and Yorkshire, wherever timber was perceived to be valuable, it was enclosed and exploited with care. There was a palpable sense of pride in Abercrombie's account of Carrick in Ayrshire:

> No Countrey is better provyded of wood, for alongst the banks of Dun, Girvan and Stincher there be great woods, but especially on Girvan whereby

they serve the neighbourhood both in Kyle and Cuninghame for timber to build countrey houses and for all the uses of husbandrie as cart, harrow, plough and barrow at very easie rates, and the sorts are birch, elder, sauch, poplar, ash, oak and hazell, and it is ordinary throughout all that countrey that every Gentleman has by his house both wood and water orchards and parkes.[35]

Similarly in Galloway, Andrew Symson spoke of the 'excellent oakwood', two or three miles long, at Monigaff, 'from whence the greatest part of the shire of Vigtoun was furnished with timber for building of houses and other uses.[36] Resources like this were not frittered away.

Nevertheless, in the period 1500–1750 woodland still declined, even in the south, especially in uplands where it was furthest from the market and where population or grazing pressure continued to be felt. In the Highlands, however, where almost all the country was upland, and which mostly lay distant from the markets and where there was relatively more wood to start with, there was a still bigger loss of woodland. This was due to a combination of human and natural causes. Let me explain.

North of the Highland line, enclosures of woodland were rare, partly because (except in Perthshire and Argyll) most of the trees were birch or, in places, Scots pine. Birch coppices only weakly – and pine not at all. Neither regenerate beneath their own shade, so there was little point in attempting enclosure and coppicing as a simple means to perpetuate the wood, which works well for oak, ash and most of the broadleaf trees further south. On the other hand, distance from the market and the relative abundance of trees did not initially make it worthwhile to proprietors to enclose the oakwoods of parts of Perthshire and Argyll either.

Woods were nevertheless valued within the Highlands. Firstly, in a bleak landscape with severe winters, they were shelter for stock: birch woods in particular enrich the ground where their leaves fall, and were seen as places of sweet grazing in spring and autumn. Secondly, they were a raw material resource for the peasant – for building, for tool-making, for fencing, for food (hazel nuts), for light (fir candles). Thirdly, in places they were indeed a source of tradable goods. Inverness had a wood market that drew birch from down Loch Ness and pine from Glen Moriston and Abernethy and Rothiemurchus; Perth, Edzell and Kirriemuir had similar markets, and the peasants in parts of Strathspey and Deeside spent much of the year cutting and carting or floating pine deals for sale elsewhere: 'The people of this parish much neglect labouring, being addicted to the wood, which leaves them poor,' said a commentator on Strathspey.[37]

Control of the exploitation of most Highland woods before the second

Fort William and Loch Eil, from Roy's Military Survey of c. 1750:
long before the clearances or the Industrial Revolution, only scattered woods
in the landscape.
British Library. Copyright reserved.

half of the eighteenth century can best be described as peasant self-regulation
modified by an element of landowner interference. The pressure of animals
on the land was in theory exercised by the soum, or ration, of a set number
of cattle, horses, sheep and goats that went with each holding. The freedom
of the tenant to take what trees he fancied from the wood was limited by the
fact that, in Scottish law, all the trees belonged to the landlord. The only
exception to this was in the very rare occasions, as at the Forest of Birse in

Aberdeenshire, where the wood was legally a commonty, and even here
after 1695 landowners could divide and privatise commons by simple
agreement.[38] The landowner's baron court fined peasants for taking wood,
but generally not so much to deter them as to raise a small income from
fines: but the process, where it operated efficiently, should have been
capable of preventing over-use.[39] Some baron courts, notably those of
the Campbells of Glenorchy in the late sixteenth and early seventeenth
centuries, carefully enjoined on tenants the duty of planting a set number
of trees every year, and of avoiding muirburn that might get out of control
and damage the woods, just as the court also busied itself with the
destruction of wolves and discouragement of poachers.[40]

Yet there was, almost certainly, more woodland decline in the seventeenth
and eighteenth centuries in the northern half of Scotland than in the
southern half or in northern England. To a limited degree this may have
been because there was more direct clearance to facilitate farming – more
assarting. In 1662 Sir Robert Gordon of Straloch recalled in Aberdeenshire
and Banffshire that woodland had been considerably more plentiful, and
that as it was cleared villages became less nucleated: 'I remember seeing
instances of this procedure in my early years. The farmers abandoned
their villages and removed each to his own possession, where any vein of
more fertile soil attracted him.'[41] Pont's rough maps of north-east Scotland
c. 1590 do not suggest vastly more wood than later, yet we need not doubt
the general point that as population rose in the sixteenth and early seventeenth centuries it put pressure on marginal areas, including wooded
ones. We should also recall Professor Dodgshon's argument that in the
Highlands as a whole population growth may well have revived in the
eighteenth century earlier than it did further south in Britain, and then did
so very strongly, creating a situation where the population had no economic
option but to press more hardly on the land.[42]

More people would mean more animals, and among the animals the
goat (everywhere particularly destructive to trees) seems to have become
increasingly numerous. No doubt the extinction of the wolf made the
keeping of all livestock easier. Increasing animal population devouring
the seedlings steadily removed the regenerative capacity of the woods.
This factor was probably much more important than the direct felling
of trees to make arable land. To quote Adam Smith again: 'Numerous
herds of cattle, when allowed to wander through the woods, though they
do not destroy the old trees, hinder any young ones from coming up, so
that in the course of a century of two the whole forest goes to ruin.'[43]

The other relevant factor is climatic. The period of the sixteenth to
eighteenth centuries spans the nadir of the Little Ice Age, particularly the

years 1500–1610 and 1670–1700, when the weather was colder and wetter in the Northern Hemisphere than at any other time in the past millennium.[44] In an area on the Atlantic rim like the Scottish Highlands, the bad weather would have magnified the ancient effects of oceanicity, of rain and gales leading to ever denser peat formation and poorer conditions for tree growth and regeneration. These effects would have been seen most severely on the north and west coasts, exposed to the most savage winds, and at high altitudes. At this period small glaciers temporarily reappeared in the Cairngorms. The upper limits of cultivation in both Dartmoor and the Lammermuir Hills in south-east Scotland descended from about 400 metres above sea level to about 200, and the upper tree line of forests in Central Europe, from the Vosges to the Sudeten mountains, also fell by 200 metres.[45]

With this combination of circumstances, that the woods in northern and western Scotland were in trouble comes as no surprise. Substantial woods disappeared, or were reduced to vestiges. In Wester Ross, for example, Timothy Pont, *c.* 1590, shows woods on the south side of Little Loch Broom and very extensive ones running inland from Gruinard to beyond Loch na Shellag: he calls them 'firwoods', alias pinewoods, which is confirmed by place name evidence. Not a trace remains today, but there is no record of external exploitation. Further south in Ardgour there are only fragments of what was once a famous pine wood in Cona Glen, twelve miles long, 'good to feed guids [cattle] therein', and in Glen Scaddle, with 'a great number of fir trees . . . there uses manie shipps to come to that countrie of Ardgoure and to be loadned with firr jests, masts and cutts'. In Glencoe and on the southern side of Loch Leven there were 'many firrwoods heirabouts', in Pont's phrase, all gone by the end of the eighteenth century. Almost adjoining them on their southern edge were the forests of Glen Orchy, of which vestiges remain at Loch Tulla. Further east but at a greater altitude was the vanished wood of Coille Mhor in Strath Errick; a more extensive wood existed than today at Glen Lui on upper Deeside, described in the 1780s as a shattered and skeletal remnant of its former self.[46]

Sometimes the decay of such woods appeared entirely natural. The Earl of Cromartie late in the seventeenth century described the gradual disappearance of an old pine wood slipping beneath the peat at Little Loch Broom, and was told by the locals that this was the usual manner in which their woods disappeared.[47] On other occasions their going appeared at first sight the fault of man. Perhaps Ardgour was overcut and overgrazed ('this Glen is very profitable to the Lord'). Or perhaps the situation there was like that in Glen Orchy when an Irish partnership bought the timber in 1722 and, according to the angry landowner, the Earl of Breadalbane, removed within three years virtually every oak tree and pine on the site. However, when the case was investigated it appeared that the Irish had kept the bargain in their written contract, which was to leave all the small and young trees under two feet in circumference at breast height. The trouble was that the Glen Orchy woods, when felled, apparently had contained no such immature trees, which implies that, previously, regeneration had failed either due to excessive grazing or to climatic deterioration, or to both. The worst that the Irish can be blamed for is that they reduced the chance of the woods to regenerate at some point in the future, but unless the climate had changed again or the peasants' stock had been excluded, the wood was presumably doomed in the long run anyway.[48]

The most telling cases of woodland decay and its consequences come from the far north of Scotland. In 1753 a tenant on the Seaforth estates in Wester Ross wrote that:

> I hear from sensible honest men that other places in the country besides my tack have now less wood and more fern and heath than formerly, so that cattle want shelter in time of storm (as we never house any) and their pasture is growing more course and scarce. I know of severall burns that in time of a sudden thaw or heavy rain are so very rapid that they carry down from the mountains heaps of stone and rubbish, which by over flowing their banks they leave upon the ground next them for a great way, and by this means my tack and other are damaged and some others more now than formerly.[49]

Increasing erosion and stream flow are just what would be expected from deforestation, in addition to deterioration of the herbiage. Even more telling were the reports from Sutherland in 1812, where the agricultural reporter told of 'a remarkable alteration on the face of this part of the country' over the previous twenty years, the widespread decay of the natural broadleaf woods with which the straths had once been covered. He was uncertain whether to blame climatic change or animal grazing, remarking that until very recently every farmer had kept a flock of twenty to eighty goats, and speaking of the 'constant browsing of black cattle' that had damaged the natural oak. Most of the wood, however, was birch. The consequence of its disappearance was serious:

> In the straths where these woods have already decayed, the ground does not yield a quarter of the grass it did when the wood covered and sheltered it. Of course the inhabitants cannot rear the usual number of cattle, as they must now house them early in winter, and feed, or rather keep them just alive, on straw; whereas in former times their cattle remained in the woods all winter, in good condition, and were ready for the market early in summer. This accounts for the number of cattle which die from starvation on these

Ancient ash and birch woodland with a rich lichen flora,
Glenkinnon Burn SSSI, Borders: the multi-stemmed ash to the left
has regrown from an old coppice stool.
Peter Wakely, English Nature.

straths, whenever the spring continues more severe than usual; and this is one argument in favour of sheep farming in this country.

Corroborative evidence for the decay of these woods was added by the minister of Kildonan (no friend of sheep farming): animals that were

formerly outwintered in the shelter of the trees until January now had to be
taken into the houses of the peasants in November, and the replacement of
'fine strong grass' by coarse heather led to a 'degeneracy of black cattle in
the parts that were formerly covered with wood'. Such were the knock-on
ecological effects of deforestation.[50]

The decline of the woods between the sixteenth and the end of the
eighteenth centuries can therefore be ascribed primarily to a combination
of poor climatic conditions and overuse of the resource for grazing and
shelter by farmers. More people led to more cattle: deteriorating weather
diminished the ability of the environment to support them, but neither
the souming system nor baron court regulations were flexible enough to
regulate the pressure. If felling by outsiders played any part at all, it was a
secondary and minor one, as with the Irish in Glen Orchy.

Where does this leave the traditional villains, the ironmasters, timber
speculators and the southerner's sheep who were certainly busy on the
Highland stage after about 1750? James Lindsay investigated the impact of
the ironworks twenty years ago, and found that they had been mostly too
small, too few and too short-lived to make much impact, but also that the
largest and longest-lived of the concerns, the Bunawe Furnace in Argyll,
needed 10,000 acres of oak coppice to keep going, and left the woods in at
least as extensive a condition when it closed in 1876 as when it opened in
1753.[51] The same could be said of the much more widespread users of
oak coppice, the tanbarkers, who operated throughout Argyll, Perthshire,
Dunbartonshire and Stirlingshire in the late eighteenth and nineteenth
centuries: they managed the coppice on long rotations of twenty to thirty
years, enclosed them and kept them sustainable for as long as their busi-
nesses were profitable, and these woods only began to go to ruin again
when imports of bark and chemical substitutes for tannin gradually took
over the market in the final two-thirds of the nineteenth century.[52] The
story in the north of England was similar. In south-west Yorkshire, coppice
management reached its climax earlier than in Scotland, in the seventeenth
and eighteenth centuries, when medieval wood pasture was very largely
converted to a skilfully managed regime of coppice-with-standards, primarily
to feed charcoal to the lead, iron and steel industries of Sheffield and the
surrounding dales. When, after 1780, coal gradually began to replace charcoal
in metal smelting, other woodland industries such as tanbark production
and the manufacture of poles simply could not sustain so much land
under coppice. By 1847 woods distant from the railways were recommended
for conversion to farmland: by 1890 coppicing in the area was virtually
finished.[53]

Coppice management was not without its environmental costs. As we have seen at Coniston, there may well have been impoverishment of the flora as a result of such intensive use. In Scotland, other sorts of trees than oak were seen as a waste of space for the coppice owner, who was urged to extirpate birch, gean, holly or willow to make way for a monoculture of oak: 'Oak and nothing but oak is the only profitable tree for coppice cuttings, and wherever such a plan is intended, nothing else should be reared.'[54] And here and there doubts were cast on the sustainability of coppicing as a crop, as at Clunie in Perthshire in 1791, where the minister noted declining productivity in the natural woodlands and attributed it partly to the ground becoming 'wasted and fatigued' by constant removal of the crop.[55]

As for the timber speculators operating in century after 1750 in the pine forests of Speyside, Deeside and elsewhere in certain Highland straths, their depredations were, indeed, at times extremely severe, especially when wars against the French gave a respite from Baltic competition. The great forests of Glenmore and Rothiemurchus were virtually clear-felled (Rothiemurchus twice), while those of Abernethy, lower Speyside and Deeside declined.[56] But often they recovered to occupy much or all of their former space, either by regenerating naturally from isolated trees and shed cones or by being planted up. Clear-felling mimics the effects of wind-throw or fire in nature, and encourages pine seed to germinate in the newly disturbed land and under the open sky.

Where the forests did not grow back on the same scale it was the fault of the ever-increasing numbers of deer and sheep in the Victorian Highlands. A pine wood likes, under natural conditions, to move its stance. That is to say, it regenerates best at the edges away from the shade of existing trees, and as the centre of an existing wood decays a new one spreads out from it. This, however, demands spare land on which to spread, and too many animals even outside the wood can prevent this happening. David Nairne gave a good account of this in 1892:

> In the beginning of the century the institution of sheep rearing on a large scale had a distinct effect upon the Highland forests. The area under wood ceased its natural expansion, the young seedlings being all eaten up, while the herbage got so rough that there was not a suitable bed for the seed to fall in.

He noted that black cattle had been more favourable to the forest as they kept the herbage down and trampled the seed into the ground so that where they had been grazing 'a luxuriant crop of trees invariably made their appearance'. How far this was true, of course, would depend on the density of stocking, as there were plenty of eighteenth-century commentators who emphasised the damage cattle could do to a wood. Nairne went on:

R. M. Adam in Glen Affric, Inverness-shire, 1930: such scatter of trees
with thick heather was probably
once characteristic of much of the Highlands.
St Andrews University Library, Adam Collection.

Then came another enemy of the woods – deer – within the last half
century. Natural reproduction can never go on in or about the forests
where deer are present, as they destroy the young trees with avidity.[57]

Overgrazing has, in the last two hundred years, continued to be by far
the biggest enemy of the wood. Studies in Deeside have shown how at
Braemar, as early as the eighteenth century, the favouring of deer by land
managers brought to a halt most of the regeneration of pine.[58]

The burgeoning forest industry also made a great impact, initially in
particular in areas like Speyside where conifers grew so well naturally. The
Seafield estates are said to have planted 31 million trees up to 1853, and in
the 1870s were sending as many as two million trees a year to the hill. By
1881, some 30,000 acres are reputed to have been planted in Strathspey
and the Rothes area. Most of these would have been Scots pine, though not
all of local provenance: some would have been larch and Norway spruce.
Much of the planting was done on land formerly bearing pine, so at this
stage it could be argued that the foresters were compensating for the
problems the native woods had in regenerating themselves.

In the twentieth century, however, and especially between 1950 and

1988, there was a vast expansion of forestry planting. The area under wood in Scotland more than quadrupled in the twentieth century, often on land bearing semi-natural woodland of one kind or another. Now the favoured crops were Sitka spruce and lodgepole pine. In Badenoch and Strathspey, half the total area of semi-natural woodland was planted up, very largely by exotic species. Only since the revision of woodland grant schemes in the last decade has government policy bent itself to expanding and safeguarding existing native woodlands and successfully encouraged the use of indigenous species in forestry plantations.[59]

This change came about because of the rising value afforded to delight. I have been concentrating so far on the economic use of woodland, but in the final section, I wish to address another theme important to their history, and certainly to their future – the pleasure (the amenity value) implicit in their presence in the countryside.

There is no doubting the antiquity of delight in the woods long before the Romantic writers of the eighteenth century: there is little trace in North Britain of the fear and dislike alluded to by Keith Thomas in *Man and the Natural World*. As far back as the twelfth century, the Yorkshire monk Walter Daniels called the environs of his beloved and remote Rievaulx secluded below its crown of hills, 'a second paradise of wooded delight' where the 'branches of lovely trees rustle and sing together' over the tumbling waters of the river.[61] In the centuries of our consideration,

Black grouse displaying: a species characteristic of the habitat above, now seriously endangered.
Neil McIntyre.

Bishop Lesley in 1578, for example, speaks of Ross 'mervellous delectable
in fair forrests, in thick woods': an account of Kintail in 1590 calls it a 'fair
and sweet countrey watered with divers rivers covered with strait glennish
woods'.[62] William Drummond of Hawthornden, c. 1620, for the first time
in literature praises the beauty of the snow on the Grampians and speaks
of 'the mountain's pride, the meadows flowery grace, the stately comlinesse
of forrests old'.[63] And in the Gaelic poetry of Alexander Macdonald in the
mid-eighteenth century, 'terms such as "woods" and "trees", along with
"fruit" and "berries" are full of emotional energy and provide the appropriate imagery of a rich countryside.'[64]

Then individual trees in the Highlands and Lowlands alike were revered
for their antiquity and association. There was the Kilmallie ash in Cameron
of Locheil's territory, 58 feet in circumference, burned by Cumberland's
troops in 1746, the bitter reprisals of an act of cultural terrorism. There
were two separate Wallace's oaks, one at Elderslie and one in the Torwood,
where the hero was supposed to have hidden: the trees were ultimately
loved to death by souvenir hunters. Some of these old icons remain, like
the Fortingall Yew in Perthshire (argued by some to be the oldest vegetation
in Europe) and the Capon oak at Jedburgh, allegedly so called because,
before the 'improvements', tenants rendered their rents of chickens in
kind beneath it.[65]

Altogether, then, there was already a well-established tradition of delight
in woods and trees to which the Romantic poets gave active impetus.
Both Burns and Wordsworth protested against what they regarded as the
senseless destruction of his woods by the profligate Duke of Queensberry.
Burns makes the sprite of the River Nith exclaim: 'the worm that gnaw'd
my bonny trees, That reptile wears a ducal crown'. Wordsworth commenced
his denunciation of woodland destruction at Neidpath in Peeblesshire with
the words, 'Degenerate Douglas! Oh, the unworthy Lord!' After this,
landowners became at least a little sensitive. When the Earl of Breadalbane
was advised by his factor to fell certain old oaks on the estate, it was with
the comment that the trees were away from the road so the tourists would
not notice.[66]

By the early nineteenth century, there is already an aesthetic critique
which placed naturally regenerated woods above plantations, and native
species above exotics. Not all contributed to it, of course, for a great age of
specimen planting of such North American trees as Wellingtonia, Douglas
fir, grand fir and noble fir, was about to begin, and large-scale forestry was
starting to be practised in Speyside and elsewhere, sometimes using Scots
pine of continental provenance but increasingly turning to exotic species as
best suited to realise a quick return from the soil.

Wordsworth praised the native oak, ash and holly and deplored the alien tree, taking his neighbours to task for solecisms like planting Scots pine in the English lakes. James Grahame of Glasgow complained not of the choice of tree but of the way it was planted. A modern conifer plantation was, he said, quite unlike a natural wood – it was dead and heavy 'one enormous, unbroken, universal mass of black' with 'something of that dreary image, that extinction of form and colour, which Milton felt from blindness'. A man might choose to hang himself there but would have difficulty in finding a single stem to which a rope could be fastened.[67] And Lord Cockburn, sober circuit judge and critic of all things modern, in 1839 visited Aviemore and condemned 'that abominable larch with which it pleased the late Rothiemurchus, as it still pleases many Highland lairds, to stiffen and blacken the land'.[68] These were the first criticisms, justified or not, of the new forestry industry.

All these examples, from Burns onwards, demonstrate the post-Romantic, post-Enlightenment conflict between use and delight which is the constant theme in the environmental history of the last two centuries. After all, the various landowners from degenerate Douglas onwards were only trying to realise a timber crop, or to plant efficiently to suit the tree species to the soil. In the eyes of their critics, however, they were destroying the natural and the beautiful which, by being in the public eye, belongs to us all.

This problem came to haunt the Forestry Commission when it was set up after the First World War. Obviously it was the job of the Commission to grow timber, and on a sufficient scale to relieve Britain of the danger of being starved of a strategic resource in the event of another war. This task was central to the first half century of its existence. But what if it was in the wrong place, or the wrong sort of woods? Almost from the start there were bitter rows between the advocates of commercial forestry and those who wanted to preserve amenity and promote outdoor recreation.[69] At the centre of dispute, as usual, was the Lake District, with Wordsworth's ghost hovering above it: eventually, in 1938, the Council for the Preservation of Rural England and the Forestry Commission agreed that an area of 300 square miles in the central zone of the Lakes should be exempted from afforestation by the Commission.[70]

After the Second World War there were similar disputes about the appropriateness of afforestation in the other new National Parks, notably in the Peak District and the North York Moors. From the 1950s there was similarly strong criticism of afforestation of the Scottish glens by Norway spruce, Sitka spruce and lodgepole pine, not least in the measured observations of W. H. Murray in his report on Highland scenery to the National Trust for Scotland in 1962.[71]

As the role of state planting declined and that of private-sector forestry companies increased, they too began to draw renewed fire from conservationists, particularly for the replacement of semi-natural woodlands by conifers: 40 per cent of Scottish native birch woods in the Highlands were lost in the decades after 1945. In the 1980s, these disputes reached a series of climaxes in the dispute over Creag Meagaidh's natural birch woods, resolved by their purchase for a National Nature Reserve; over the Abernethy Caledonian pine forest, resolved by their grant-aided purchase by the RSPB; and finally by the dispute over planting the Caithness and Sutherland Flow Country which ended in the Solomonic judgement by the Secretary of State dividing the Flows half for the forest industry and half for birds. In 1988, Nigel Lawson, urged on by the *Daily Telegraph*, pulled the rug out in his budget by removing most of the tax advantages for the forest companies. It was a limited victory for delight over a form of use which seemed, to the Conservative government of the day, less and less to justify the public money that supported it.

But it should not be imagined that the Forestry Commission and the forest industry were ever entirely on the side of use over delight. Even in the early days the Commission was anxious to promote Forest Parks and to allow direct access to public woodlands, to the extent that the Treasury in the 1930s expressed a fear that an agency intended by Parliament to plant trees was becoming an agency to promote hiking.[72] In 1944, its chairman, Sir Robert Robinson, expressed amazement over the Dower Report on National Parks, asking why people should want to walk in open country when it was so much nicer to walk in woodland.[73] Things were made easier for the Commission when in 1958 national forest policy was formally amended to include social objectives, and when in 1963 public recreation and amenity were admitted as desirable functions. Their enthusiasm for public use of the woods, and the success of the eight Forest Parks in existence by 1955 (four were in northern Britain) stood the Commission in good stead. When, under the last government, privatisation was proposed and rejected, it found it had substantial support from a public who feared that its amenities for delight would get short shrift in the private sector.

Furthermore, the Forestry Commission ultimately accepted that it had a responsibility for conserving ancient and semi-natural woodlands. The use of the term 'ancient woodland' to identify woods that had been occupying a site since at least 1600, and which were distinguished from 'recent woodland' by ecological features such as a rich flora and invertebrate life, had originally been introduced by Oliver Rackham in 1971. The concept was subsequently refined and developed by him and by George Peterken

for the Nature Conservancy Council. As late as 1978 the Forestry
Commission firmly rejected the concept, but, prodded by a Select Committee
of the House of Lords and persuaded by an academic conference at
Loughborough in 1982, it changed its mind. By 1992 it was accepting the
NCC's Ancient Woodland Inventory to identify ancient woods, providing
guidelines for their management and grants for their maintenance. In
twenty years 'ancient woodland' had moved from academic concept to
acceptable heritage and become a national institution.[74]

So if the woods of imagination were the stuff of patriotism and Romantic
contemplation, the woods of reality have been contested ground. The battle
is not over. The Woodland Trust in particular campaigns for better protection, as most ancient woods still have no statutory protection and many
are destroyed every year. But I think we see the contest lessening, use and
delight beginning to become reconciled, a peace process well begun.

CHAPTER 3

MAKING AND USING
THE SOIL

Good fertile soil is a precondition for our existence on earth. On it depends the crops necessary for our use, and the vegetation of the woods and moorlands in which we take our relaxation and delight. As an environment for farming, European soils appear at first sight remarkably robust, unlike for example the shallow and friable soils of South Australia, of parts of tropical Africa and of the dustbowl states of America. But this is often an illusion. They are neither unchanging in the face of natural processes nor immune from damage by human use. One theme in this chapter is the recurring problem of sustainability in using the soil. Another is the extent to which the soil and its contents are a human artefact.

The fragility of soil even in northern Britain must have been learned repeatedly from an early date. To take but one example, when in the third century before Christ, farmers in the Scottish Borders felled the natural woodland of the Bowmont valley to open the land to cereals, it led to an extraordinary erosion event in which the soil washed down the river and came to be deposited as a band of coarse, silty sand, 8 cm thick, on the floor of Yetholm Loch: a reminder that unsustainable use is not necessarily a monopoly of modern times. Whether in an attempt to provide adequate soil depth or to hinder further erosion, or both, the prehistoric farmers then adopted the practice of running their rigs in narrow terraces parallel to the hill, reminiscent of how in the American mid-west farmers learned in the face of the dust bowl to adopt contour ploughing. In the Iron Age, prior to the arrival of the Romans, the valley was supporting a population at least as large as any that has been there since.[1]

What the farmer needs from soil is two qualities, good structure and content, but those of modern agricultural soils are, to a degree seldom appreciated, consequences of past human action as well as of natural endowment. The structure must allow the free formation of roots from the germinating seeds. This can be assisted by drainage, the addition of humus to light soils and of marl to heavy ones, and by the constant action of implements and tools. The content must include the right chemicals in the right form to allow the plants to absorb them as nutrients. The farmer today can buy them in bags, but that option was not open to his forefathers. The

64

critical nutrients are nitrogen, phosphorus and potassium. The main problem was probably nitrogen, since the others form comparatively stable and plentiful compounds and may not have restricted increases in output until after the nitrogen bottleneck had been overcome, around 1840. It has recently been argued that a tendency to export phosphorus in grain and animal products from medieval farms in Oxfordshire was enough to account for declining yields on the manor of Cuxham in the early fourteenth century,[2] but on most lowland soils, at least, unless very calcareous, as few were in northern Britain, the reservoir of phosphorous in the soil should have been so large that small net exports would have made no difference.

Nitrogen is a different matter, since it only occurs in forms suitable to be taken up by plants in unstable compounds readily dissolved or oxidised once forest or grassland cover is disturbed. The earth may have begun rich, but it wore out. By the middle of the nineteenth century, according to Robert Shiel, 'much of the old arable land in Britain appears to have lost two-thirds of the soil nitrogen which was present before farming began'.[3] The agricultural history of Western Europe can be written in terms of efforts to plug nature's leaking cornucopia with sufficient manure. Or, as they put it in Yorkshire in 1794, 'it is upon the solid foundation of manuring that every good system of husbandry must be built'. More colloquially, muck is brass, and farmers in Scotland removed their hats as a mark of respect before a well-made midden.[4]

In very wet areas, like northern Britain since the Bronze Age, rain creates a further problem by leaching out nutrients and encouraging the formation of podsols where the topsoil becomes progressively more acid. Cereals grow best in soil that is approximately neutral, with a pH level of about 7; if it drops to 4, the level on ordinary moor or bog, many agricultural plants cease to grow altogether; anything below that is progressively fatal to other plants as well.[5] The effect of low pH is to lower the availability of nutrients, and to allow metals which can effect plant growth (aluminium, iron and manganese) to become mobile.

Furthermore, the atmosphere contains both nitrogen-based and sulphur-based gases on which the rain acts. These occur naturally, but a very important feature of the last century and a half has been the increase of nitrogen and sulphur oxides as a result of global industrial pollution. In rain, nitrogen is naturally deposited to form compounds immediately accessible to crops, but at recent unnatural levels this may encourage grasses on moorland at the expense of heathers adapted to less fertilised ground. Similarly sulphur is deposited in rain as sulphuric acid, falling on the soil and eating away the stonework of Pictish monuments and Gothic churches. In combination with the nitrogen it changes the once soot-black, recently cleaned, Victorian buildings of Manchester and Glasgow to a modern algae-green.

In short, there is a myriad of ways in which human activity, from the Iron Age to the jet age, has altered the soil: and even without our intervention it is not quite the fixed and unchanging element in our environment that we often take it to be.

A decline in the productivity of English manors in the thirteenth and fourteenth centuries due to nutrient loss has been considered a possibility since Postan raised the matter thirty years ago.[6] In 1994, the Danish environmental historian Thorkild Kjærgaard similarly argued that Denmark early in the eighteenth century had faced a major ecological crisis that hinged on the failing availability of nitrogen in the soil.[7] It was not merely that nitrogen was hard to utilise through contemporary agrarian practices, so yields were low and agriculture in many places stagnant: that was already an accepted area of discussion among agricultural historians of early modern Europe. His point was that output per acre was significantly *declining* in Denmark, not that it was stagnant. Due to the extent of recent deforestation, the spread of sand dunes and the consequent rise of the water level, nitrogen was either leached out, buried or locked up by waterlogging and acidification.

There are several objections that can be raised to Kjærgaard's thesis. The extent of deforestation, for example, was by his figures a decline from 25 per cent to 12 per cent of the land cover, 1550–1750, which was perhaps insufficient to have created the hydrological disruption alleged, and the extent of sand-blow only affected 5 per cent of the country, which left 95 per cent unaffected. But he produced much circumstantial evidence both of declining standards of living and of agrarian decay, and his book at least raises the question as to how far farmers in other parts of Europe might actually have been unable to plug a leak in the cornucopia, perhaps not everywhere of recent origin, but a long-term one that had been progressing since the start of farming. Perhaps it occurred in cyclical form, serious when population grew and more land was opened to ploughing and leaching, ameliorated when population fell and nutrients accumulated in permanent pasture.

Was Scotland like Denmark? It was certainly deforested to a greater degree, though not recently in the eighteenth century: bogs and rain had been a problem since antiquity. But the Little Ice Age was reaching its nadir in the seventeenth century, accompanied by increasing wind, colder temperatures and more precipitation. Furthermore, population growth between 1550 and 1650, which was plainly significant though hard to quantify, must have been accommodated largely by resuming cultivation on

"'Tis almost incredible how much of the mountains they plough':
medieval cultivation rigs on Arthur's Seat, Edinburgh, unploughed
since enclosure as a royal park in the mid-sixteenth century.
Royal Commission on the Ancient and Historical Monuments of Scotland.
Crown copyright.

marginal land of the uplands abandoned since the famines and plagues of the fourteenth century, and thus the sixteenth-century farmer cashed in the short-term bonus of two hundred years of nitrogen and phosphorus accumulated in the turf. Once it was used up or washed out in a century of deteriorating weather, might there not indeed have been a renewed crisis of nutrient availability in northern Britain?

The idea is unproven without better statistical evidence of crop yield trends, but it is not completely fanciful. Except on clays and 'marblie' black soils, said Andrew Garden, writing of Buchan in 1683:

> Much ground is now with often ploughing and manuring, turned so thin that it is altogether useless either for grass or corns, and because of this many men's estates are not able to keep up the ancient rentall.[8]

Population in Scotland, as in England, ceased to grow after about 1650, but in Scotland stagnation was followed by a crash in the famines of the 1690s, possibly leading to a 13 per cent fall in numbers. Through the century, the rate of emigration from Scotland was perhaps the highest in Europe, at times reaching one in five of adult males. There was a widespread decline in the consumption of meat and other animal foods, which in most societies would be seen as a drop in the standard of living. Real wages dropped in Scotland, both in town and country, after 1650, and did not improve again for about a century.[9]

It looks like a society unable to cope with the challenge of providing very adequately for its members even by its own previous comparatively meagre standards, and as 90 per cent of the population lived in the country-side, the likeliest source of the problem was in faltering agriculture, which is where many contemporaries were inclined to place it. Northern England was better off, but perhaps not much. Seventeenth-century fiscal lists seem to show that the seven poorest counties of England were those of the north, and the total assessed wealth of the area beyond the Humber, one-fifth of England, sometimes scarcely equalled that of Wiltshire. Taxation historians wisely urge us not to take such evidence at its face value: in the north, tax-evasion was rife, institutionalised, large-scale and of long standing. Yet had there not been a large regional disparity already, the London government would not have let the north get away with it for so long.[10] 'To pass from the borders of Scotland into Northumberland', wrote a respondent in the *Statistical Account*, 'was rather like going into another parish than into another kingdom.'[11]

If there was a crisis of this kind, why was it not equally evident in southern Britain? There could be at least four reasons. Firstly, the south was more benign both in climatic terms and in the proportion of the land that was lowland and of soil naturally high in pH values. Secondly, farmers there had perhaps got a better grip of nitrogen provision through wider use of leguminous crops that fix nitrogen through nodules in the soil – not necessarily clover, which was known but not widespread yet even in East Anglia, but peas, beans and sainfoin. Thirdly, the widespread practice of enclosure organised the deposit of dung more rationally than by allowing animals to drop it on the open hill. Finally, better rotation of fallows and leys facilitated the use of nitrogen accumulated in pasture more efficiently.

If this is true, it follows that the environmental crisis in the north might have been averted or reversed by copying or adapting best southern English practice relating to legumes, enclosure and rotation, which indeed is what Scottish Improvers were advocating from the late seventeenth century onwards. Broadly speaking, the solution, not only to averting crisis but to raising agricultural productivity well above all previous levels, came eventually in Scotland, as it did in Denmark and across the whole of Northern Europe, through widespread use of nitrogen-fixing crops, above all red clover, and also through using turnips to enable more stock to overwinter in yards and greatly to increase the supply of dung for the cereals. All this is too well known to need enlarging upon here. In the case of northern Britain, the treatment of acid soils by lime and marl to raise the pH level was also of outstanding importance, the 'principal way of gooding the soil', as it was called in the Lothians in the seventeenth century, the 'first and most important step in the new system of husbandry', as it was described in Roxburghshire in the eighteenth.[12]

It is worth emphasising that the scale and completeness of change in Scotland between 1750 and 1830 certainly deserves the *sobriquet* Agricultural Revolution, whatever doubts may linger about the term in England. The latest careful examination by Professor Devine concludes 'the overall impression is that oat yields in Angus and Lanark in the 1790s were triple seventeenth-century averages and in Ayr and Fife approximately double.'[13] By the beginning of the nineteenth century the Lothians were treading on Norfolk's heels for the reputation of being the most advanced and productive agricultural region in Great Britain. Visitors were coming from as far afield as Poland and the United States to talk with the agronomists of Edinburgh University and stay on farms like Fenton Barns near Dunbar to examine the latest rotations that kept the land productive and in good heart.

It was a revolution based on making soil. What has been insufficiently emphasised is the lengths that farmers went to in order to modify its physical character by the transport of heavy materials from one place to another. They had no notion of modern science, but they discovered what worked. Liming worked. Thus James Robertson's report to the Board of Agriculture in 1794 on south Perthshire observed:

> There is nothing more common, and perhaps few things more difficult to be accounted for than, when lime is spread on short heath, or other barren ground, which has a dry bottom, to see white clover, and daisies, rising spontaneously and plentifully, the second or third spring thereafter, where not a vestige of either, nor even a blade of grass, was to be seen before.[14]

His colleagues agreed. The reporter in Northumberland said you could tell 'to an inch, by the superior verdure' where the grass had been limed and the white clover arose, and in north Yorkshire that 'three or four chaldrons of lime per acre . . . entirely change the natural produce' of moorland over-run with ling or bent 'to that of a fine turf, full of white clover'.[15]

They also learned from their mistakes. Sir Robert Sibbald in the 1690s had already described lime as the usual modern way to improve the land, but said it also sometimes ended by sterilising the fields and killing the fish in the streams if done too lavishly.[16] Again a century later a Berwickshire minister in the *Statistical Account* said that the 'burning of the fields' with excessive lime emptied the streams of trout and salmon.[17]

Despite such trial-and-error approaches, the agricultural soils of northern Britain are precious fixed capital for modern farmers, and in many cases they have been made as valuable as they are today by the enormous effort of their predecessors. Of the two qualities needed from soil, the nutrient content may now come mainly from a factory, but the all-important structure is a more lasting gift from history. Quite apart from their economic value, some of these soils are so remarkable a cultural and historical artefact that they have as good a case for preservation as most scheduled archaeological sites.

Take, for example, the coastal township of Marwick in West Mainland, Orkney, examined by Professor Donald Davidson. Here deep topsoils, of a type known as 'plaggen' found also in the Netherlands, Denmark and elsewhere in Western Europe, survive still, exceptionally rich in phosphorus and over a metre thick. It can be shown that they began to be constructed in the late twelfth or early thirteenth centuries and continued to be built over 700 years, until at the start of the present century new inorganic fertilisers provided a way of feeding the land with less effort. The topsoils at Marwick were formed by transporting almost 200,000 cubic metres of material from the hill to the steadings in the form of stripped grassy turves which were used as animal bedding and then, soaked in dung and urine, transferred to the arable land. Seaweed was also laid on the fields to top up the turves. These soils were constructed on those parts of the arable area called in Orkney the 'tounmal' that were kept in continuous cultivation by a single occupier. They grew that form of single-rowed barley known as 'bere'. The soil in the area beyond, called 'the tounland', was periodically redistributed among neighbouring farmers and sometimes rested from cultivation, but was nothing like so well manured. At Papa Stour on Shetland comparable soils were made. Here all the arable topsoil on the tounland was deepened, not just the part nearest the steading, perhaps because there was less expectation than in Marwick of a tenant enjoying a particular spot over a length of time.[18]

These examples from the Northern Isles were not affected by the Agricultural Revolution in any basic way. From the high Middle Ages to the early twentieth century on such remote spots on islands and Highland coasts the farmers made soil as they had always done. It was different, however, in the fertile Lowlands between the Humber and the Moray Firth, or between the Dee and the Clyde. Here the eighteenth century was an age of experiment and of information-sharing. Manure was a topic that even

Misusing the soil: erosion gully in a field of linseed near Yetholm,
Borders Region, after three days of torrential rain, March 1992.
Donald Davidson.

the most genteel felt free to discuss, as it had been liberally treated by classical authors from Virgil to Palladius. James Donaldson in 1697, wrote of
what was under discussion in Scotland in his time:

> Some curious persons recommend horn and hoof, blood and guts of cattle,
> and shells of fishes, and salt-petre etc., as very strong and durable nour-
> ishment for the ground; yet seeing these cannot be had by everyone, I shall
> speak of those which may be had everywhere, VIZ. dung of cattle, ashes,
> lime, marl and sea-ware.[19]

All these commodities and more were actively in use in the eighteenth century. 'Mr Rudd of Marsk makes composts of kelp ashes, seaweed, slam [a refuse from the alum works], and lime', said the reporters of the *General View of the Agriculture of the North Riding* in 1794, 'all mixed together with earth and finds them to answer well.' Others in Yorkshire made use of 'the cleanings of ditches and sometimes the shovellings of roads, mixed with lime'. In Northumberland the *General View* reported on coal ashes, 'chiefly used in the vicinity of the principal towns, as a dressing for grass land; for this purpose they are found of considerable benefit, especially upon strong, coarse and wet lands.' In Ayrshire, the reporters described the use of waste from soap manufacture, 'the earthy part of kelp and barilla, mixed with lime'. In Clydesdale the farmers used (with animal dung) horn shavings, woollen rags, parings of leather, peat and coal.[20] On the coast of Aberdeenshire they used mussels from the local shores, an 'excellent manure', whale blubber from the Greenland fleet which raised 'a very rich crop for the first year, and a tolerable one for the second'. Mr Auldjo of Aberdeen bought dogfish in bulk once their fish oil had been extracted by local merchants, and mixed them with horse dung and spent tanbark.[21]

Great effort and care was often put into manufacturing the various composts. Mr Auldjo, for example, was also one of many who used peat ash. He took a large quantity of peat from the moss, set it alight, added a layer of clay, another of turf and another of clay until the heap was as high as a man could reach, obtaining a thousand cart-loads at a single burning: sixty loads for an acre of oats or peas did wonders, and thirty to forty as a top dressing for grass 'never fail to mend the crop'.[22] In Caithness, the name of Lord Meadowbank was given to a system of composting unburnt peat and seaweed in similar layers, six inches upon six inches, until the heap was five feet high and tapering on the top like a house: it was completely turned over in March or April and used to fertilise bere in May or turnips in June.[23] In Yorkshire, Isaac Letham of Barton, near Malton, made yet another such layer-cake, alternating thirty bushels of earth to eight bushels of soot or pigeon's dung in six- inch bands: 'he has found it answer well for wheat and artificial grasses if laid on early in the spring.'[24]

One of the advantages of these composts was that, like legume cultivation, it improved the nutritional status of the ground without multiplying the weeds, whereas manuring with farmyard manure spread the undigested weed seeds over the fields with the nutrients, and great effort had to be expended on controlling them. Thus although making the composts was very labour intensive in winter and spring, it might actually save labour later in the year.

Probably nothing had greater potential for permanently changing the face of the countryside than lime and marl, the latter a generic term that covered shell deposits in clay or any soft rock such as mud or silt stone with a high calcareous content. We have already seen how they had the capability of altering both the chemistry and structure of the soil. They turned counties like Roxburghshire and Berwickshire, once, in the words of a writer of 1784, a patchwork of 'good land . . . checkered with moors and other barren ground', into their modern regular and uniform pattern of fertile fields. But the labour involved was immense. One farmer in the area reported using 25–30 cart-loads of shell marl per acre, another 150–200, a third recommended 450–600 cart-loads of clay or rock marl per acre. Even if some carts carried more than others, each acre was transformed by an extraordinary input of human and animal muscle.[25] There was no ambiguity about the quantities involved when the agricultural reporter for the East Riding of Yorkshire recommended adding wold chalk stone (with similar properties to marl) to 'strong fresh land, in the proportion of sixty tons to an acre'.[26]

Because soil-making inevitably involved carrying large quantities of heavy material of low value, it was seldom taken far. The modern traveller in Fife can see how the soil is darkest nearest the shore, reflecting the intensity of former applications of seaweed, and in Shetland, the diminutive breed of ponies, whose only work was to carry the seaweed, did not take it beyond a narrow coastal strip.[27] Human ordure from the cities and towns was carried into the neighbouring parishes, but seldom further afield. Edinburgh's dung increased in popularity in the eighteenth century, which made the city cleaner. The price went up according to the minister of Duddingston from 2d. a load in the 1730s to as high ('in some circumstances') as 1s. 6d. a load in the 1790s.[28] It passed toll-free on the roads, and when in 1787 the turnpike operators tried to charge for it the farmers organised a boycott and Edinburgh had to try to dispose of its own sewage.[29] Such inexhaustible supplies of cheap nitrogen kept the parishes round about Edinburgh (Cramond, Corstorphine, Liberton, Duddingston, North Leith) under rotations with more wheat and less fallow than was customary, but it clearly did not pay to take the manure beyond this inner ring. Where there was access to river or canal it might go further. Hull sent its stable dung, street sweepings and night soil by water to many parts of the West Riding, delivered on board at 4s. a load and sold to the farmer at 5s. to 7s., depending on distance.[30]

The commodity that, under favourable circumstances, was sent furthest afield was manufactured lime. One of the most remarkable undertakings was that of the Earl of Elgin at Charlestown in West Fife, where coal and lime exceptionally occurred together within immediate reach of the sea.

Investing £14,000 in 1777–8 in mines, waggonways, kilns and quays, he quarried 80,000 to 90,000 tons of limestone a year, sold partly as unburnt stone and partly as manufactured lime, selling the latter for shipment in 1,300 cargoes annually to as far away as the Moray Firth (150 miles by sea): it was not uncommon to see 40–50 vessels waiting to load during the summer.[31] Arguably this works generated more shipping than the entire annual maritime trade of Scotland a century before. This, though, was the exception that proves the rule, which is that few soil-making materials were moved many miles: it was a matter of recycling or repositioning nutrients within a narrow radius around a community.

Although the intention was always to enrich the land, it did not always succeed. The effects of over-liming were severe but fortunately short lived. Much controversy arose out of the widespread practice of taking in land from the moor for arable cultivation by paring and burning, which involved stripping peat or turf from rough pasture, burning it and mixing the ashes with the underlying soil before ploughing for a cereal crop. The immediate returns were often high, but the long-term effects disappointing or sometimes downright harmful. In 1794, the West Riding agricultural reporters summed up much opinion when they said that their information on the matter was 'various and contradictory. . . upon the whole, it is a practice that should be gently used, as it tends in a material degree to exhaust and impoverish the soil.'[32] Presumably it removed any humus above a podsol, and after the initial benefit of the nutrient-rich ash had been exhausted, the underlying soil was left permanently more barren than if humus and nutrients had been ploughed in at the time of the original reclamation.

Very similar, but older and much commoner, was the harm routinely done to moor and pasture by stripping turf as bedding for stock and thatch for houses, or directly for fertiliser for arable land, as earlier described at Marwick and Papa Stour but in fact general through the length and breadth of Scotland in the older farming system. 'Grant it as a good manure', said Mackintosh of Borlum, one of the early Improvers, in 1729, 'I think it is a very dear one, to tir [strip] one area to put upon another.' He described it as leaving nothing but pebbles and gravel 'a mile in circumference in many estates'.[33] It created what is known as 'skinned land', recognised as a common feature to this day over wide areas of, for instance, the Isle of Lewis.

Transporting materials from one part of a farm or township to another, of course, changed the nature of the soil for better and for worse in different places. Professor Dodgshon in a series of important articles has shown how rising population in the central and western Highlands in the eighteenth century led to increasing effort to improve cereal yields by labour-intensive nutrient transfers of this sort.[34] The disadvantages were, firstly, that it was

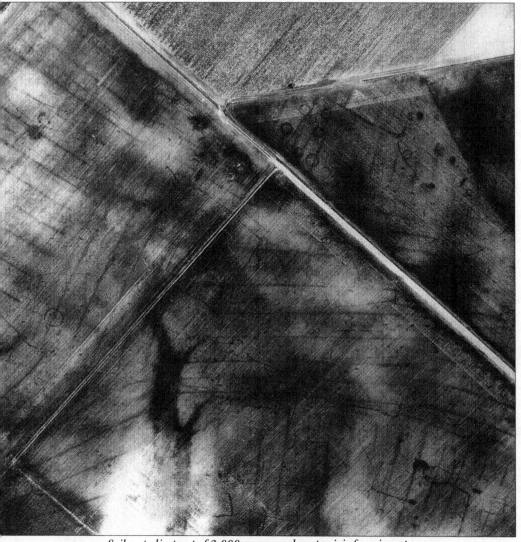

Soil as palimpsest of 3,000 years: modern prairie farming at
Leuchars, Fife, 1978 with Bronze Age hut circles and medieval or
later rigs showing as crop marks.
Royal Commission on the Ancient and Historical Monuments of Scotland.
Crown copyright.

unlikely that increased output per acre would be matched by increased
output per head, and, secondly, it undermined the capacity of the moor to
keep the manure-making and food-producing animals on which the townships also depended.

It is easy to see the increasing popularity of potatoes as a solution to this problem, along the lines that Ester Boserup proposed in her studies of the relationship between demographic pressure and technological change on the land.[35] Under population pressure, a supposedly conservative peasantry switched with alacrity to a hitherto little known crop that would yield far more nutrition per well manured acre than the old cereals. Nutrient transfer had to continue in making the lazy beds for the new crop, and back-breaking work was involved for the women in bringing seaweed and other manures to the beds, but potatoes completely altered the equation between effort and reward.

Self-sufficiency of local areas in soil-making persisted for a long time. As late as 1843, Falkner's *Muck Manual*, a most helpful *vade-mecum*, had little in it that would not have been instantly familiar generations before. It dealt with the excrements of horses, cattle, pigs, sheep and people, with bones, fish and fish oil. What it called 'mineral and artificial manures' were sea-weed, lime, gypsum, ashes and burnt clay. The only portents of the future were a reference to Liebig's work on the chemical components of manures, followed by a report on an experiment by the Morayshire Farmers' Club to raise turnips by means of sulphuric acid and bone dust, and a reference to 'experiments on guano' brought from Peru.[36]

Yet the *Muck Manual* was written on the threshold of a change of epoch-making consequences: agriculture was about to be released for ever from the need to depend on local resources for improving the soil. In the following decades, the Victorian farmer came to benefit from a potent mix of the railways and the steamships, free trade, informal empire and science. The import of bones increased so fast after Liebig's experiments that the great German chemist worried about the transfer of phosphorus and accused Britain of hanging 'like a vampire . . . on the neck of Europe'.[37] The height of the guano islands off Peru was lowered by more than fifty metres as seabird excrement deposited over millennia in a dry climate was mined off by Chinese coolies and shipped to the fields of Britain: some reckoned it thirty times as effective as farmyard manure. Geological deposits of phosphates and nitrates were dug out in suitable spots across the world from Cambridgeshire to Chile: the latter had its heyday between the 1880s and the 1920s. As late as 1925 more than 30 per cent of the world's fixed nitrogen supply came from Chile, where it was mined with British capital. And as if this wasn't enough, there were substantial imports of animal feedstuffs such as maize, Middle Eastern beans and locust pods, and oilseed cake that was a by-product of the cotton fields of America and the flax fields of Eastern Europe. The imported feeds made nitrogen-rich dung that, mingled in the straw of stall-fed cattle, was particularly excellent as conditioner on heavy clays. Nutrient transfers were now between continents, not just between one field and the next. Sustainability at home now rested upon access to the fertility of the Third World.

At the same time, advances in ploughing techniques were underway. In Scotland, the replacement of the heavy old ox-drawn ploughs of tradition by light swing ploughs pulled by a pair of horses had been an eighteenth-century advance, but gradually the ploughs cut deeper until, if required, the subsoil itself could be turned up. Stephens, writing in 1846, spoke of the old ploughs achieving a 'mere skimming of about four inches of soil' that had created 'an effete powder by constant cropping', below which was a thin slaty compressed crust, and below that again 'the black virgin mould remained untouched'.[38] His description suggests an exhausted topsoil holding an impoverished soil biota, and the ability of the Victorians to cut into untapped sources of fertility as though on a newly discovered continent. But once tapped, of course, as in America or Australia, such fertility had to be maintained rather than mined.

There are one or two suggestions in the history of birds that as a result of all this attention the topsoil indeed became much wormier and richer in invertebrate life than hitherto. One species that experienced an explosion in numbers and distribution was the starling. It is a species that appears to have been present in northern Britain in the eighteenth century and then suddenly to have disappeared. It reappeared equally suddenly before the middle of the nineteenth century. The Duke of Argyll spoke of his 'great interest' when in 1837 he saw his first, outside a posting inn at Northallerton, on his way south. By 1844 it was reported as 'very common' in Yorkshire, but by 1907 as having increased enormously over the previous half century and as having pressed up into the dales. Meanwhile it had invaded Lowland Scotland and spread throughout: 'the increase of this species was enormous and rapid'.[39]

Late Victorian and Edwardian naturalists frequently remarked on a general increase in songbirds in their lifetime, attributing this sometimes to control of hawks and corvids. Scientists today, however, insist that there is no correlation between numbers of small birds and those of, for example, sparrowhawks and magpies, despite a widespread popular belief to the contrary. The 'songbirds' in question are likely to have been thrushes and blackbirds, perhaps also robins and skylarks, all of which, like rooks and starlings, depend on worms, beetles, grubs and other invertebrae in the soil. Intensive organic farming, as most of this was, could hardly fail to multiply this fauna.

Although the heyday of heavily capitalised, high-input, high-output farming ended abruptly around 1875 in the face of cheap American grain, dependence on imported fertilisers and feedstuffs (becoming cheaper with every decade) increased rather than declined. They were supplemented now by chemical by-products of other industries, such as nitrogen compounds produced in the manufacture of coke and coal gas, and basic slag, rich in phosphate, from the ores prepared for the Gilchrist-Thomas steel furnaces. If grain prices were falling for farmers so were the costs of maintaining fertility, and since the price for animal products (except wool and mutton) usually fell less than the price of wheat and barley, the situation was tolerable if they could avoid over-dependence on corn.

Consequently, except for hill sheep farms, the years between 1875 and the First World War were not a depression in northern Britain in quite the same way as in the corn-fixated south. Dairy production and the supply of quality meat to the urban market combined with cheaper nutrient inputs left a decent profit margin to farmers who could take advantage of them. Depression only bit here in the quite different circumstances of the inter-war years.

This period after 1875 was, however, also one in which understanding of the soil itself made great strides, informed by a series of distinguished scientists from J. B. Lawes at Rothamsted in the 1870s and 1880s to R. G. Stapledon at Aberystwyth in the interwar years. One point of clear understanding to emerge was that nutrients could be equally well supplied in artificial as in natural form, though some manures (like guano or ammonia salts) would provide fertility for one year only, while others (notably farmyard manure) would have an effect throughout a rotation and beyond. Robert Elliot of Clifton Park in Roxburghshire carried out important trials on his estate in the 1880s that emphasised the critical significance of soil structure – the physical condition of the soil, its suitability for root development and power to retain warmth and moisture. He was strongly in favour of an emphasis on clover and farmyard manure, kept in rotations of four years under grass and four under subsoil ploughing. Elliot was an inspiration to Stapledon, whose advocacy of ley farming and reseeding the edge of the uplands with wild white clover and cocksfoot grass became a major influence on husbandry in northern Britain before the Second World War.[40]

By the end of the nineteenth century, farmers had become used to the calculations of science. The *Country Gentleman's Catalogue* of 1894 used a different approach from the old *Muck Manual*, informing its readers that one ton of farmyard manure would contain 9–15 lb of nitrogen, the same of potash and 4–9 lb of phosphoric acid; that in six months an animal would void six tons of manure in a closed yard onto two tons of litter; and that for wheat, barley or oats five tons of farmyard manure was required to replenish the ingredients abstracted from the soil per acre of each crop. The list of manures was much longer, with ten sorts of guano listed, as well as super-phosphates made out of guano or bone ash, nitrate of soda, gas lime, gypsum, wood soot and peat moss – an amalgam of the old and new in which the organic still remained more prominent than the artificial.[41]

With the Second World War the economic and political status of farming was transformed, first by blockade shortages and by the provision of guaranteed markets and assured prices in the Agriculture Act of 1947. Enormous increases in land productivity followed. It is worth emphasising the scale. Comparing British agriculture between 1936 and 1986, wheat output rose more than ninefold, barley by thirteenfold; oats (in 1936 the grain most widely grown) dropped to a quarter of its former level, but grain crops overall rose by about sixfold. Beef production rose by 60 per cent, egg production by 90 per cent. The income of farmers in real terms doubled between 1936 and 1976, but was dropping back towards 1930s levels by 1986 and has since fallen lower.[42]

These UK averages conceal large regional variations, and the barley barons of East Anglia or the Lothians came out of it better than the marginal land farmers of much of northern Britain. Disaster now stares the latter in the face. By 1998, average net farm income on Scottish hill farms was under £6,000 a year, less than £2.80 per hour for a 40-hour week and a third of what it had been two years before. It can hardly be disputed that the tremendous postwar transformation of the productive capacity of the land benefited many enterprising and energetic individuals who brought it about. It may be disputed, however, whether it permanently benefited either the farmworkers, who even in the 1970s were not getting their fair share of a rising standard of living, or the farming community as a whole in the long run.[43]

The increases in output from the soil were achieved by capital-intensive changes made affordable by public subsidy. This took many forms. In 1936 there were a million working horses on British farms; in 1946 there were 436,000, by 1960 only 46,000 and for the years since no figures have been available. By 1986 there were half a million tractors instead – and more than half of those were capable of over 50 horsepower each.[44] Farm size doubled and the farm labour force fell by two-thirds over the same period.

Nothing was more significant, however, than the complete chemicalisation of both soil inputs and pest and weed control. What had in the days of the agricultural revolution been shifted only a few miles, and in Victorian times delivered by steam power between continents, was now delivered to the farm in a bag from the factories of the specialist giants of agrochemicals, ICI and Fisons, and later Monsanto. Pests that had been previously

Hay meadow at Harker's House SSSI, Swaledale, Yorks, 1986: a seriously
endangered habitat in the uplands.
Peter Wakely, English Nature.

ignored or destroyed by hand were eliminated by a spray. The change had
started before the war, but it was only after 1945 that the chemical industry
truly gripped agriculture. In the half-century to 1980, there was a twenty-
fold increase in nitrogen usage on British farms, a 2.5-fold increase in
phosphorus and a sixfold increase in potassium (until 1974 both nitrogen
and phosphorus fertilisers received subsidies).[45] In 1940 there were 1,100
tractor-mounted sprayers in England and Wales; by 1981 there were over
74,000.[46] In 1944 only sixty-three products were approved for use by
farmers as pesticides; in 1976, there were 819. In 1956 these were made
up of thirty-seven types of chemicals; by 1985 the number of approved
chemicals had risen to 199. More varieties may have meant safer and
better-tested ones after the DDT scare of the early 1960s, but it certainly
did not mean smaller usage. The tonnage of active ingredients in insecticides used on farms in England and Wales more than doubled during
the 1970s alone.[47] With renewed recession, 'targeted inputs' and 'precision
farming' are now becoming watchwords, but for thirty years it was 'drench

and be sure'. And now a new danger is appearing with genetic modification, that herbicide-resistant crops will give farmers power to remove all other flora from the fields, and insect-resistant crops with built-in pesticides will enable them to wipe out all insect life.

Chemical farming on this scale was a complete novelty. The question of its full impact is too big to address in full in a limited space. I shall restrict myself to three points. How has it affected the soil itself other than by making it more productive? What, broadly, has been the effect on biodiversity? Has it had wider implications for the fortunes of agriculture?

The adverse effects of the application of chemical fertilisers and pesticide residues on the soil itself have long been suspected, and little attention has been paid in the last fifty years to what so troubled previous generations of husbandmen, the maintenance of good soil structure. In the opinion of some soil scientists, there is a danger of taking the work of past generations too much for granted.[48] In 1971 L. B. Powell put it strongly. 'The notion that the soil is an inert mass which needs only a constant plastering with chemicals to give increasing yields', he wrote, is 'one of the tragic follies of the twentieth century'.[49]

Take the problem of soil erosion. Under continuous cereal cropping and using only chemical fertilisers, the levels of organic matter in soil, up to 6 per cent under established woodland and 12 per cent under old pasture,

Corncrake: a species of traditional hay meadows reduced to final refuge in the Hebrides.
Laurie Campbell.

eventually fall to 2 per cent.[50] Light soils with such low organic matter and with low clay content readily become unstable, liable to wind blow or winter flood gullying, especially if lying in large, modern fields devoid of windbreaks: between the 1940s and the 1970s the length of hedgerow declined in Scotland by over a third, by the 1980s by over a half,[51] and the height to which existing hedges were cut greatly reduced what efficiency they had to resist the wind. Fields are particularly at risk in northern Britain when the land is cultivated for winter-sown cereals and potatoes, creating wide, bare expanses which wind and water can rip apart; further south, the cereals germinate in time to cover the ground before the onset of winter and spring storms, and there has actually been reduced soil erosion since winter cereals became popular in the 1980s.

How significant winter erosion is in Scotland is difficult to quantify, but it affects some of the best soils in the Moray Firth, Fife, Lothian and Berwickshire, and losses have been put in places as high as 15–45 tons per hectare.[52] Foresters in the Lowlands have even suggested that they do not need to apply fertilisers to their woods, since the wind already blows enough nitrogen, potassium and phosphorus from surrounding fields.[53] Erosion is a novel problem on most farms: apart from sand-blow on the machair and at the Culbin sands and parts of Northumberland, it has apparently never occurred on any scale since the original forest clearance in northern Britain. It is part and parcel of the agricultural changes that chemicalisation made possible. On the other hand, when it reaches serious proportions farmers readily have the solution in their own hands by abandoning as unsustainable (in the north) winter-sown cereals on sensitive soils.

Then there is the question of the direct effect of chemicals on the character of the soil. In 1968 a wet summer was followed by widespread autumn waterlogging and a failure to produce a tilth for sowing next year's corn: to some it seemed an indication that continuous cropping might have irrevocably damaged the soil itself. The Agricultural Advisory Council, however, concluded that even at 2 per cent organic content the soil was perfectly viable and that the problem lay in neglected field drains.[54] Though heavy machinery has been implicated not only in crushing the drains but in physically compacting some of the clay soils, this problem is less serious in northern Britain where clays are relatively less common.

The comforting initial conclusion is that chemicalisation has not had an irreversible or general adverse effect on the soil itself, though in the judgement of many it has made its products less palatable, even less nutritious and healthy, and the application of so much nitrogen, phosphorus and pesticide has certainly had an effect on water quality in streams, rivers, lakes and wells. On the other hand, it is premature to be complacent about soil structure: firstly, because soils are known to be, in many cases, inherently unstable entities, subject under a critical load of stress to changing suddenly and unpredictably;[55] secondly, we know alarmingly little about the effect of chemicals on soil biota, and what knock-on effects that may have on soil ecology and hence structure.[56]

When the question arises of the effects of chemicalisation on biodiversity in the last half century, there is less room for doubt. The widespread use of pesticides was a novelty: the word is not even found in the *Concise Oxford Dictionary* of 1950. Initially of particular value to farming were the chlorinated hydrocarbons, including DDT, aldrin, dieldrin and heptachlor, which had the unfortunate incidental qualities of being extremely persistent in the soil (and elsewhere in the environment) and of being soluble in animal fat. Starting as seed dressings to protect grain or sugar beet against fungal or insect pests, they moved up the food chain into the partridges or pigeons that ate the corn and into the foxes or falcons that ate the birds, with deadlier concentrations at each step.[57]

As early as 1928 the RSPB had advised against the unwise use of toxic chemicals.[58] In 1945, one or two scientists warned against indiscriminate use of pesticides in the environment. By the late 1950s sportsmen were beginning to notice their effects: Wentworth Day's uncompromisingly titled book *Poison on the Land: The War on Wildlife and Some Remedies* (1957) was concerned about dwindling game birds. In 1960 the Master of the Fox Hounds Association reported 1,300 fox deaths in eastern England. They were thought to be due to some strange new disease. In 1961, the Nature Conservancy, which had been becoming increasingly concerned about pesticides for a decade, launched an enquiry under Derek Ratcliffe that proved the link between chlorinated hydrocarbons and breeding failure in peregrine falcons, and so demonstrated contamination of the food chain.[59] The average thickness of the eggshells of peregrines fell by a fifth after 1947, and of sparrowhawks by a quarter, so that the eggs broke in the nest as they were being brooded. In 1962, Rachel Carson in America wrote *Silent Spring* which also brought popular anxiety about pesticide abuse to boiling point in Europe.

The government in the UK felt the prick of the political spur and responded with commendable speed, despite a fight by the chemical industry which tried to discredit the work of the Nature Conservancy in much the same way as the tobacco lobby took on medical scientists over cancer research.[60] In two stages, in 1964 and 1969, the main persistent organochlorines were withdrawn from most normal agricultural uses, although it was not until 1981 that dieldrin and DDT were finally banned in all but the most exceptional circumstances. Peregrines, which by 1964 had fallen to 44 per cent of their prewar numbers and were then only breeding successfully in central Scotland, recovered to their previous levels within twenty years, as did other raptors like kestrels and sparrowhawks. Some prey species that had been affected, like stock doves, showed a similar capacity to recover.[61]

It would be reassuring to think that that was the end of the story, but in terms of the overall effect of modern chemical farming on biodiversity, it was only the most dramatic first chapter, exceptional in its happy ending. There are too many indicators, qualitative and quantitative, that point in another direction. The atlases of the British Trust for Ornithology show the distribution of breeding birds in Britain in 1968–72 and again twenty years later: between the two, the corncrake, for example, contracted its range by 76 per cent, the corn bunting by 32 per cent. Their common bird census, held since 1971 (after the banning of chlorinated hydrocarbons), now covers more than twenty-five years: in that time such typical farmland birds of northern Britain as grey partridges declined in number by 80 per cent, corn buntings by 77 per cent, tree sparrows by 95 per cent, skylarks by 58 per cent, linnets by 53 per cent, and so on (see Figure 3.1). To translate these percentages into actual numbers, more than a million skylarks have vanished from the British countryside in the last two decades.[62] It is true that these figures are calculated on a UK basis from data biased towards south-east England, where declines have apparently been more serious than in the north, since less of the land surface is affected by intensive farming. But even a county like Fife, that still enjoys a healthy population of such species as grey partridges and linnets, has experienced the overall trend, the distribution of corn buntings having contracted by two-thirds since 1970, and the numbers of singing males in the remaining population having dropped by over 20 per cent to about ninety-five individuals in the last six years.[63]

In the nineteenth century corn buntings were so numerous, pulling at straws in the stackyards, that infuriated farmers took their revenge by eating corn-bunting pie. Today a party of a few score, in any area where they are still considered relatively common, is worth a note in the local bird report. More generally, in the nineteenth century, seed-eating birds must have been staggeringly common. Between 1815 and 1820, the churchwardens of Godshill on the Isle of Wight paid out each year for an average of over 10,000 'sparrows' (probably denoting a medley of finches, buntings and sparrows) destroyed as pests in their parish. Sparrow clubs devoted to the same purpose and killing comparable numbers existed at least in England until the interwar years.[64]

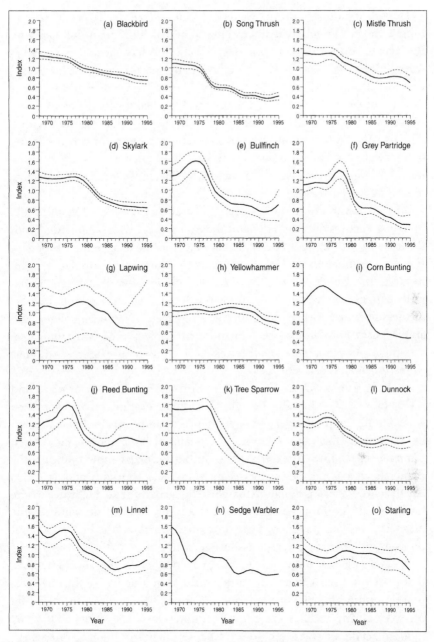

*Figure 3.1 Decline in farmland birds. Decline in fifteen species of farmland birds
in Britain as measured by BTO Common Bird Census 1968–94. A 'smoothed
Mountford index' shows the extent of the decline (except for two species)
between 95 per cent confidence intervals.*
Source: *British Ecological Society, Siriwardena* et al.
Journal of Applied Ecology, *vol. 35 (1998), pp. 24–43.*

Despite the demise of the organochlorines, since the 1970s the use of fungicides and insecticides in cereal farming has more than doubled and, due also to sophisticated herbicides, cereal crops have become so clean of weeds and insects as to be very inhospitable places for most birds (see Figure 3.2). It could get worse: genetically modified crops could result, as in parts of the USA today, in a countryside virtually wiped free of sustenance for birds. But even now the causes of decline go wider than pesticides. The use of so much fertiliser helps to establish patterns of cereal farming that leave few winter stubbles and so little food after the harvest, and the need for large fields in which to wield large machines leads to the decline in nesting sites in spring, and so forth.

The history of the starling is again illustrative. Between 1956 and 1966 breeding success fell markedly and then recovered, which coincided with the use and withdrawal of persistent organochlorine insecticides. Breeding success continued to improve until about 1980, when the population again began to plunge. Between 1972 and 1997 the number of starlings halved in Britain. Grass was converted to arable, fungicides reduced the richness of the soil biota, spring ploughing was replaced by autumn sowing and arable–pasture rotations were abandoned in favour of constant cereal cropping. As feeders on leatherjackets and other invertebrae, starlings suffered from all these changes.[65]

What is true of biodiversity among birds is equally true of plant and invertebrate life. Though the same quantitative studies have not been made for all groups, the decline in wildflowers and bees is well documented by scientists. Take, for example, the beautiful large yellow and brown bumblebee, *Bombus distinguendis*. Distribution maps show how, like the corncrake, it has vanished in fifty years from lowland Britain, and now, also like the corncrake, it is common mainly in those parts of the Hebrides where low intensity crofting has survived as the only way to farm the fragile machair. The same march of agriculture that wiped out the bird wiped out the bee, and eradicated plants associated with both.[66]

It is, of course, a commonplace that an older generation always believes that the world, particularly the rural world, is getting worse. In this case, however, unfortunately for the farmers, a great phalanx of people over fifty can very clearly recall a more lovely countryside than today's. Of many modern accounts of last childhood Edens, it is worth quoting one which has the merit of being by a scientist, who was both knowledgeable about and sympathetic to the farmer's business. This was Keith Mellanby's world:

> As a boy I knew Upper Teesdale well. Above Middleton-in-Teesdale there
> was little arable, but wild flowers, of which the spring gentian is the most

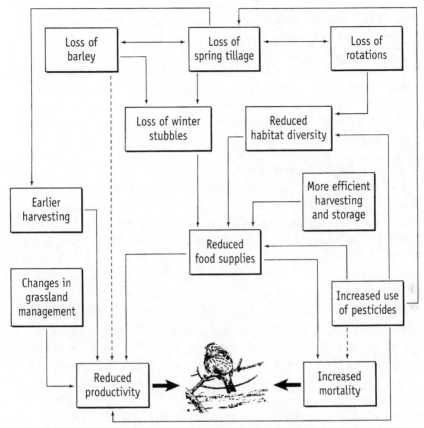

Figure 3.2 The decline of the corn bunting.
Source: *English Nature.*

famous, abounded in suitable localities. On the damper grazing meadows
the globe flower was common in early summer. The mealy primrose grew
freely on drier ground. In 1979 I went on a farm walk on fields I knew fifty
years ago. The management of the grass was excellent. Cattle and sheep
grazed at a density which would have been unthinkable in the past, and
hay and silage of the highest quality was made from different fields. Careful
use of lime and basic slag, with the return of all manure to the soil, ensured
a steady rise in fertility and the continuation of top quality pasture. The
soil fauna clearly flourished under the grass. But the beautiful flowers I
remembered so well had all disappeared.[67]

Here, as so often in the history of twentieth-century agriculture, was a clash
between use and delight. But it was not a new one. Here is an eyewitness
account of Fenton Barns, in mid-nineteenth-century East Lothian:

This great farm almost reached the sublime. It went like clockwork. Its
fields, of from 20 to 30 acres, were all rectangular. There were no odd cor-
ners, no thickets, no hedgerow trees, no ragged, any-shaped pastures.
The quickset hedges were clipped and low and narrow like those of a
garden. No wild rose or old man's beard rambled on them, no may or
blackthorn blossom lit them up, neither did the violet or the primrose find
a lodging beneath their shade. There were no open ditches, and the
plough ran right up to the roots of the fence. The land was as clean as a
well-kept garden.[68]

It was an ecological desert already.

Using the soil, of course, always alters and historically has often acci-
dentally enriched biodiversity. The very corncrakes and corn buntings,
linnets and partridges whose decline is now lamented were the unintended
beneficiaries of early arable farming, with its grubby soils and weedy fields.
The starlings and sparrows profited from rising agricultural productivity in
the nineteenth century. The gentians, globe flowers and mealy primroses
grew in upland hay meadows. None of these would have found Britain a
hospitable place covered with forest and bog. Some of the plants now sorely
missed by botanists, like the corn-marigold or 'gool', were regarded in their
day as major pests of good husbandry: 'Gool days' were organised in
Scotland in the eighteenth century in a vain attempt to eradicate it from
corn fields, just as the sparrow clubs existed to keep down bird pests.

Finally, has there been a wider significance? Modern definitions of
sustainability involve the need to carry the local population with you in all
that is done to modify or to conserve the environment. Part of the problem
for late twentieth-century farming has been, as well as doubts over public
health in an age of intensive farming, the speed and scale of biodiversity loss,
of the triumph of use over delight, leaving a public angry and concerned at
the extent to which the remembered countryside has been swept aside by
subsidy and spray. The farming community was no longer seen as sons of
the soil, but as a body of businessmen and lobbyists, quick to arrogate to
themselves the title of guardians of the countryside but most of them doing
little to deserve it. Herod might as well have been a patron of the Save the
Children Fund. Thus when farmers needed urban friends, as they did
when the BSE crisis hit in the mid-1990s, they were nowhere to be found.
The common perception, however oversimplified, was that by misuse of
nature the farmers had brought catastrophe upon themselves: perhaps
the public forgot too quickly how their own votes had supported and per-
petuated a policy to which the farmers merely – if eagerly – responded. In
the twentieth-century conflict between use and delight, the chemicalisation
of farming gave agriculture a triumph, but may it yet prove to have been a

hollow one for the farmers concerned. In this way, as in others, questions about its sustainability can no longer be dismissed as scaremongering. They loom large and threateningly over the industry.

CHAPTER 4

COMMANDING THE WATERS

In the ancient Middle East, from which so much of European culture is derived, water was precious and rare. The Persian word from which we derive paradise described the pleasure gardens of Mesopotamia where water was the source of life and flowed from a fountain to the points of the compass in four channels, as we can see symbolically represented in the unique survival of an English medieval cloister garden at Westminster Abbey.[1] For a chosen people dying in the desert, Moses miraculously struck the rock to produce the water that preserved them. For the Christian, the water of baptism in its stone font stood at once for the River Jordan and the promise of eternal life for the saved. These are symbols of scarcity and value to which we all react with instinctive recognition.

Drought-stricken lands use water sparingly and regard it reverentially. But here on the north-western edge of Europe the reality is that water was, and is, superabundant. Even within Britain, however, there is a striking difference between north and west, and south and east, and with global warming it is one that is likely to grow. Between 1941 and 1970, for example, average rainfall in Scotland was more than twice that of the territories of Thames Water and Anglia Water, and residual rainfall or run-off (i.e. total rainfall less what is returned in evaporation either directly or by plants) was 75 per cent in Scotland, compared to 30 per cent of a much lower total in Thames and 25 per cent in Anglia. Since 1970, there has been a marked increase in rainfall and run-off in northern Britain associated with mild wet winters, while there has been a tendency for drought in the south.[2] In many parts of the world (for example, Spain and California) conflicts between water-rich and water-poor regions are a source of angry political tension. It is not impossible that, with further climate change in the next century, northern Britain and southern England could find themselves similarly at odds. Certainly in our area, historically, water has been considered a nuisance more than a resource, except where it was a resource for power. It was not a treasure unless it had been rarified. Uisgebeatha – whisky, the water of life – came not from a fountain or a holy stoup but from a bottle.

Over the course of four hundred years, the most striking thing that we have done to water in northern Britain is to reorder it. Once it has fallen from the sky, it lies in quite different places and behaves in quite different ways from what it once did. Water in the fields hardly rests on the surface today unless the storm has been exceptionally severe, and then only briefly. Then there were innumerably more little ponds and marshes, and the curving furrows between the rigs which formed the only field drainage were often permanently clogged with rushes. Rivers that engineers have tried to confine between banks and levees then meandered wildly and changed course frequently, leaving oxbow lakes, islands and alder swamps along their courses: Roy's military survey of *c.* 1750 showed twenty-one islands on the river Tummel between Pitlochry and the Tay confluence – only four are detectable on aerial photographs today (see Figure 4.1).[3] Such works were seldom an unqualified success. As early as 1846 it became evident that flood prevention measures on the Tweed had made the upper reaches more flashy, not less. And on the lower stretches of powerful rivers the floods continue, sometimes with devastating effect as when the Derwent, the Tay or the Cart burst their bonds and tear through

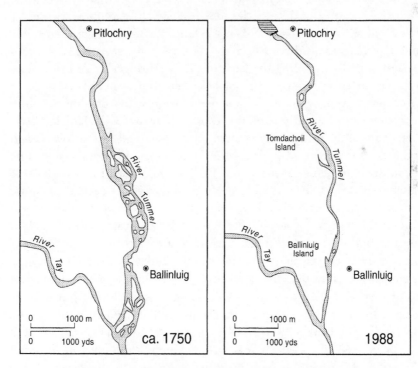

Figure 4.1 Channel changes on the River Tummel, Perthshire.
A wide wandering gravel-bed river becomes one with a much narrower single course in 1988.
Source: *Gilvear and Winterbottom (1998), p. 98.*

housing estates that planners have foolishly allowed in their old territories. The last major flood on the Tay, one of the rivers most strait-jacketed with embankments, was in 1993; it was the twentieth in 200 years and caused an estimated £28.5 million worth of damage.

The past, semi-permanent occupation of the valleys by water pushed cultivation up the slopes, which is why not only prehistoric hut circles but medieval and later field systems are common in the uplands. The Revd Thomas Morer, a London clergyman serving as chaplain with the army in Scotland in 1689, described it with southern eyes:

> They have many fine vallies almost useless, on account of the frequent bogs and waters in such places . . . and 'tis almost incredible how much of the mountains they plough, where the declensions, I had almost said the precipices, are such, that to our thinking, it puts 'em to a greater difficulty and charge to carry on their work than they need be at in draining the vallies.[4]

The spur to riverine embankment and arterial drainage was agricultural improvement. It is hard to recall today how much there once was under water. There are many thousands of hectares of what is now prime arable land, especially in northern England, that were in the seventeenth century fen and mire. Considering the fame of the Cambridgeshire, Lincolnshire and Norfolk fenlands, it is surprising how their equivalent in Yorkshire and Lancashire have evaporated from general memory. In Yorkshire south of the confluence of the Ouse and the Trent, 70,000 acres of Hatfield Chase were 'constantly inundated' before Vermuyden and his fellow Dutch undertakers commenced to drain it in 1626, in an enterprise eclipsed by his subsequent achievement at the Bedford Levels in East Anglia (see Figure 4.2). At its heart was Thorne Mere, 'almost a mile over', the historic bed of which is a modern bone of conservation contention over peat extraction. Potterick Carr, 4,000 acres near Doncaster which fell to Smeaton and his engineers after a private Act of Parliament in 1764, was one of many outliers generically known as the Yorkshire carrs.[5] The valleys of Holderness were fens that opened on to salt marsh, and Hull was an island surrounded by brackish water: as early as 1402 the Julian Dyke brought fresh water in by aqueduct, but by 1597 it had collapsed and the town could get drinking water only by lighter, 'to the excessive charges of the inhabitants'.[6] The Vale of York was summer pasture overlying clay, full of meres and flooded in winter. Here the Derwent in particular was liable to flood at any time of year – in Leland's words 'this ryver at greate raynes rageth and overflowith', and in Defoe's, two centuries later, it 'overflows its banks and all the neighbouring meadows always after rain'.[7] On the other

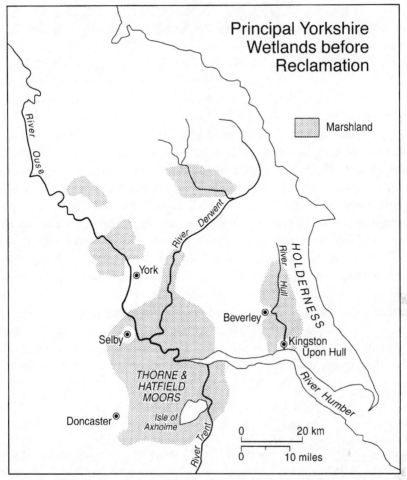

Figure 4.2 Principal Yorkshire wetlands before reclamation.

side of the Pennines there were a series of meres in the Lancashire coastal plain between the rivers Douglas and Alt, of which the largest was Martin Mere, judged by Leland to be four miles long and three miles wide, covering perhaps 6,000 acres and partially drained in 1697, more thoroughly a century later.[8] Nor were such features confined to the North of England. Dumfriesshire and Ayrshire, Fife and Berwickshire, and the lands along the Moray Firth, for example, all had extensive marshes and lochs which have either vanished or diminished almost to vanishing point. Loch Leven in Kinross-shire, still the largest natural sheet of water in Lowland Scotland, was reduced by one-third in surface area early in the nineteenth century to provide additional arable land.

The history of the bittern in northern Britain tells us much about the former distribution of the extensive reedbeds and standing shallow waters on which it entirely depends. In the British Isles today it is seriously endangered, with fewer than two dozen pairs left, mainly in East Anglia, but still with some as far north as Leighton Moss in Lancashire. In the seventeenth and eighteenth centuries it was very widespread. In Scotland, in Anglian times, the settlement of Bemersyde in Berwickshire was apparently named for the boomer in the marsh. In the sixteenth century, the 'Bittour' was regular quarry for royal hawking parties in their forays across the Lowlands. In the seventeenth century, it occurs among a list of the birds even as far north as Sutherland and was described as 'making a great sound in the summer evenings and mornings' at Lochar Moss in Dumfriesshire in the Borders. By the 1790s it was still reported from Loch Leven, from Dumfriesshire and Ayrshire, but had become 'very scarce' at Alloa on the Forth and at Saughtree in Roxburghshire. By the 1830s it had been almost entirely driven out by drainage from Dumfriesshire, Kircudbrightshire and Ayrshire, and lost from Billie Mire in Berwickshire, from Alloa and from the Strathmore marshes in Angus. In Argyll in 1843 the minister of Inveraray said that 'the bittern, which old people remember as common forty years ago, has utterly forsaken the parish.' It had gone from Scotland by 1850.[9] A similar story could be told of Yorkshire, where it was once abundant enough to have its own vernacular name and to be the subject of folk rhymes:

> When on Potteric Carr the Butter Bumps cry
> The women of Bulby say summer is nigh.

Smeaton's engineers put a stop to that nonsense. By the start of the present century little more could be said of the Yorkshire bittern than that old people remembered it having occurred near Beverley.[10]

It was, of course, in the nature of improvers to describe the great watery places as wastes, and the value of the land when reclaimed was indeed often great. Cobbet in 1830 described the ground won from the brackish marshes of the Humber, as, with the exception of the Cambridgeshire fenland, the richest and most fertile ground that he had seen in the whole of England, and the value of the land at Hatfield Chase was raised from 6d. to 10s. an acre by the Dutch adventurers. But such figures obscure the worth of the place under water to those who lived there, for the marshes were, in their own terms, incredibly productive, and attempts to drain them might be met, as at Hatfield, by active and violent resistance by the inhabitants.[11]

The fowler was among the busiest of the marshland occupations. Among the documents in Leland's *Collectanea* is an account of what was purportedly served at the enthronement feast of the Archbishop of York in

1466. Even allowing for an element of medieval exaggeration, it sounds like conspicuous consumption at its most gross: 400 swans, 2,000 geese, 3,000 mallard and teal, 204 cranes, 204 bitterns, 400 herons, 400 plover, 200 dozen ruff, 400 woodcock, 100 curlew, 4,000 pigeons, 104 peacocks, 200 pheasants, 500 partridges, 100 dozen quail and 1,000 'egrittes'.[12] The peacocks, of course, would have been domesticated, as perhaps would have been the pigeons; the partridges, the quail and the woodcock would have been birds of farmland or wood. But all the remainder were marsh birds.

The 'egrittes' may raise the eyebrows of the Yorkshire birdwatcher today, as there are few modern records of any species of egret in northern Britain. Dr W. R. P. Bourne, however, has recently reviewed evidence that suggests that not only egrets but three other species of heron now vanished from the breeding avifauna (the night heron, the purple heron and the little bittern) survived on the English wetlands until after 1600. Turner in 1544 reported seeing 'white herons' in English heronries; at least one English medieval illuminated manuscript portraying birds familiar to the artist shows an egret; in recent years great egrets have started to breed on the Dutch polders and little egrets have become common visitors to the southern English coast. Perhaps they were quite abundant on these medieval wetlands.[13]

Apart from the egrets, the species list is further confirmed and even extended by the household book of the Neville family: in 1512 it listed cranes, herons, snipe, bittern, quail, larks, dotterel and bustards (the last two no doubt from the Yorkshire Wolds). In 1526 a banquet at the marriage of Sir John Neville's daughter included peacocks, cranes and bitterns; in 1530 another daughter's marriage feast served up cranes, herons and bitterns. In 1528 he acted as Sheriff, and his charges for the Lammas assizes included twelve spoonbills at a shilling apiece and ten bitterns at 13 shillings and 4 pence.[14]

Cranes and spoonbills have been extinct as breeding birds in Britain for three hundred years, but the ruff, for example, continued to breed on the remnants of Hatfield Chase until the 1820s. Pennant in 1766 described a regular business of taking ruff in nets, fattening them in captivity and selling them for the table for 2s. or so each.[15] With the introduction of duck decoys on the Dutch model in the seventeenth century, many thousands of wildfowl were taken in them annually in the Yorkshire marshes, at Martin Mere in Lancashire and in many other places. They ceased operation about 1800.

Nor were birds the only produce. Fish, at least at Martin Mere, were still more valuable, and here and in Yorkshire the marshes provided reeds and

*Humberhead Peatlands National Nature Reserve, 1996: a conserved
remnant of the formerly immense Yorkshire wetlands around Thorne Moors.
Peter Wakely, English Nature.*

rushes for the floors of cottages, thatching and candle-making, peat for fuel,
brushwood from the carrs for fuel and light constructional work. Above all
they provided, in their drier parts, valuable common pasture for cattle in
summer, the loss of which was the main source of dispute between the
drainage undertakers and the local peasantry. But there were also disputes
between the graziers and the fowlers: as early as 1570 on the Leconfield
estate of the Earl of Northumberland it was said that the 'dryft of the
cattell dyd disturbe the bredying of the wyld fowle and especially of the
wyld swanns'.[16] One does not get in England or Scotland, however, as one
does in France, land set aside by the landlord for wildfowling and thus
preserved from subsequent reclamation: wildfowling in Britain never had
the social cachet or mass upper-class following of grouse shooting, and
remained a plebeian profession, or at least sport whose lone-wolf followers
carried little weight against the forces of improvement.

It was not only in the great marshes and meres of Yorkshire and
Lancashire that wetlands had economic value for local people. It was the
same the length and breadth of northern Britain. In Scotland, for example,
bog-hay from the broad undrained straths was an important source of
winter fodder for stock: at its best a bog-hay meadow, a tangle of wild
flowers, sedges and grass, could yield between 12.5 and 19 cwt of hay per
acre. What was said to be the largest in Scotland was the Carron Bog in

Stirlingshire, four miles long and a minimum of one mile wide. It now lies almost entirely beneath the Carron Reservoir, but in its heyday in the late eighteenth century was described by the local minister as adding 'great liveliness and beauty to the general face of the country. The scene it exhibits during the months of July and August, of twenty or thirty different groups of people employed in haymaking, is certainly very cheerful.' Like much about the use of watery places, it had a communal aspect to its use which added to the resentment felt about its transformation, after draining, into private property.[17]

The great reordering of water was the triumph of the water engineers, often spectacularly so as in Vermuyden's draining of Hatfield Chase or in the Victorian dam-building feats, of which more below. But arguably even more important was the achievement of the little-known names who perfected field drainage, for in doing so they altered the micro-ecology of every field in the kingdom. Before the nineteenth century various methods of field drainage were attempted, apart from the simple rig and furrow where the water partially drained on the surface along the furrows. A modification of this was practised on the heavy clays of the Carse of Gowrie as late as the 1840s, with open ditches in a herring-bone pattern carefully maintained by spades. Henry Stephens in 1846 told about coming across in his own undertakings in Scotland 'the drains which our forefathers made in loamy soils', placed on retentive subsoil below the

Leatham Moss, Central Region, Scotland 1998: the wetlands erased.
Copyright Lorne Gill, Scottish National Heritage.

reach of the plough (which did not cut deeper than four inches): these
were loose stone tunnels, readily blocked by moles and silt, and liable then
to burst out as springs to produce 'the very mischief it was intended to
remedy'.[18] A better, traditional southern English alternative was a trench,
two feet or so deep, lined with branches of alder, thorn or heath (or broom
in a Scottish modification) and back-filled with soil. This was introduced
into the Scottish Borders before the end of the eighteenth century.[19] These
may have helped a little, but the endemic environmental problem in
northern Britain was well put by an Aberdeenshire minister writing in
1842 about the experience of generations of his own farming family:

> The stagnation of water on the low ground utterly precluded tillage, while
> the arable lands were over-run with noxious weeds, and chilled from
> November to May by innumerable land springs.[20]

The real breakthrough in field drainage involved three steps. The first
was Joseph Elkington's method, for which he received a prize from the
Board of Agriculture in 1797, of dealing with underground springs by
carefully locating them and draining them with deep trenching and con-
duits constructed six feet or more below the surface of the ground.[21]
Elkington was a Warwickshire farmer, and this worked best where rainfall
was moderate and water from springs the main cause of soggy ground. In
Scotland, his method seldom sufficed because there was so much more
surface water, often saturating the loam over an impervious subsoil. The
next step came in 1831 with James Smith of Deanston, a cotton mill owner
and farmer from Stirlingshire, whose system of 'thorough drainage'
entailed laying many more drains in parallel lines 16 to 21 feet apart and
$2^1/_2$ feet deep, accompanied by the first subsoil plough that broke the hard
pan.[22] It was better attuned to the needs of the wet north. Thirdly, good
field drainage demanded cheap and durable drains. Earthenware horseshoe tiles were introduced from Staffordshire into Cumberland around
1819 by Sir James Graham of Netherby, perhaps the first of many
landowners to build tileworks on their own estates. John Reade's cylindrical
clay pipe manufactured by Thomas Scragg's pipe-making machine of
1845 was more effective than the horseshoe, and favourable loans were
made available from government from 1846 to help landowners drain their
estates. It was the earliest of all agricultural subsidies apart from the Corn
Laws themselves.[23]

As a result, the early Victorian period saw enormous input of labour,
horsepower and materials directed to drain the fields on a scale that we can
barely begin to imagine, but with consequences that transformed landscape,

ecological conditions and agriculture. Henry Stephens described the estate of Yester, not in some marginal area but in the heart of fertile East Lothian. Before drainage:

> The ridges were gathered-up high on all the farms, and rushes grew luxuriantly in every open furrow, while the entire surface was wet and poachy throughout the greater part of the year. Upon a soil in such a state thorn-hedges did not thrive . . . Pieces of waste land were found in every field.

After 'thorough-draining' and subsoil trench ploughing, the fields became dry, readily pulverised, uniform, hedged and tidy: they harvested sooner and yielded more, often growing wheat where before only oats had been possible, and yielding up to twice the weight of grain per acre.[24]

Meanwhile, water engineers everywhere continued to drain the marshes, big and small, and to turn them into arable land. The drama was perhaps most evident in Yorkshire. Thomas Allen in 1828 praised the Act of Parliament of 1811 for enclosing the still-extensive commons of Hatfield and Thorne and converting 212,000 acres of 'the wide extent of waste' into 'waving fields of corn'. There was much else to report in the East Riding:

> Within the last half century the vast commons of Wallinfen and Bishopsoil, containing upwards of nine thousand acres, have been enclosed and cultivated, besides several others of inferior extent: and a vast and dreary waste, full of swamps and broken grounds, which in foggy or stormy weather could not be crossed without danger, is now covered with well-built farm houses and intersected in various directions with roads.[25]

It was the same the length and breadth of northern Britain. As the Victorian sportsman Charles St John observed of the Morayshire plain in 1847, 'the best part of this ground for wildfowl is gradually getting drained, and what was (a few years since) a dreary waste of marsh and swamp has now become a range of smiling cornland'. But the reaction of his keeper, 'taking a long and sonorous pinch of snuff' when he discovered that his favourite haunt for ducks and geese had become a field of oats, was to observe, 'Well, well, the whole country is spoilt with their improvements, as they ca' them. It will no' be fit for a Christian man to live in much longer.'[26]

It has been estimated that the nineteenth century saw the drainage of five million hectares of the English lowlands.[27] Field drainage reached an apogee between 1840 and 1875, but when the price of grain fell subsequently many considered the returns on a thorough system of under-drainage uneconomic. However desirable it might be on permanent pasture, it was explained in 1894, 'there is not time for the land nearly to repay the outlay before it requires draining again.'[28] Consequently between 1875 and 1940 there was a certain regression towards boggy fields, which no doubt favoured breeding snipe, redshank, mallard and teal, and frogs and orchids and so forth, that were characteristic of small-scale wetlands. Almost all of this land, and a great deal more besides, was reclaimed for cultivation during, and especially after, the Second World War, and handsome grants became available both from the UK government and under the Common Agricultural Policy to enable farmers to drain their fields afresh, and to mop up remaining wet areas that had defied the Victorians, who, for all their energy, lacked modern draglines and plastic piping.

One reason for keeping ponds was to water cattle: very many small waters of this sort have been filled in or become lost by vegetative succession. They have become redundant over the last half century as piped water is now supplied for animals at a trough, a gain for animal welfare if a loss for biodiversity.

The effect on biodiversity on farms has of course been serious, most measurably on populations of wading birds. The situation was worse in southern England, but in Scotland snipe have disappeared from 18 per cent and redshank from 24 per cent of the areas where they bred even as recently as 1970, but most of the losses must have come long before than that.[29] The impact has been no less on wetland flora, insects and amphibians. Dr Croucher of York University is studying the plight of the water spider, *Argyroneta aquatica*, which must be typical of that of many creatures similarly placed. It is a widely distributed species of standing fresh water with no relatives in its genus, genetically uniform presumably because water was once everywhere and transport between populations so easy. It exists now in a much drier landscape only in scattered populations that must have difficulty in communicating with each other, and in modern fragmented populations its genetic viability may become uncertain.

The uniformity of modern farms is a far cry from the mosaic of wet and dry ground in each field described by commentators before 1850. On the other hand, a halt was eventually called to ambitious and expensive arterial drainage schemes to reclaim the remaining major northern marshes and riverine wetlands. In 1950, for instance the recommendations of the Duncan Committee for a subsidy from the state to drain completely Lochar Moss in Dumfriesshire and the upper parts of Strathspey were rejected by the government.[30] This did not stop J. M. Bannerman, Liberal Party politician and luminary of the influential Scottish Peat and Land Development Association, from calling (around 1962) for the installation of pumping machinery in flat river straths such as Strathspey, where 'scores of thousands of acres' awaited reclamation, and for the wholesale reduction of loch levels by removing silted sandbanks at the outlet: Loch Lomond could be reduced by four feet.[31]

Then there came a distinct change in attitudes. In the 1970s, the Royal Society for the Protection of Birds purchased the partly drained Insh marshes in Strathspey, and by blocking the drains began to return them to their older condition as a habitat for scarce birds, rare invertebrates and interesting plants. Local farmers are still puzzled and resentful that land with agricultural potential could be treated in this way. In the winters of 1988/9 and the following year flooding downstream led to renewed calls for major engineering on the Spey that might have endangered the marshes and destroyed the geomorphological interest of the untamed Feshie, a torrential tributary river system unlike anything north of the Alps. The Secretary of State found against the engineering solution after a public inquiry, on the grounds that the usefulness of the proposed works was small in relation to the potential environmental damage. It seemed like a small but significant victory of delight over use. The Insh marshes now hold more breeding waders than the Norfolk Broadland.

So far we have spoken of the drainage of waters, but no less interesting a feature of the past two centuries has been the creation of new ones. In southern Britain today, probably the most frequently encountered new water bodies are gravel pits, created by extracting materials to make roads and concrete, such as those that stretch for miles along the valley of the Thames around Oxford. There are few of these in northern Britain, though there are some (for example in the Howe of Fife), and mineral extraction of another kind has in the north created its own new wetlands through subsidence following coal mining, as in the Clyde Valley round Hamilton, Loch Ore in Fife, the Yorkshire Ings and even a revitalised though diminutive Potterick Carr without its original butter bumps. Nevertheless they all provide vital new watery habitats, sometimes in what has otherwise often become a very desiccated countryside.

All the large and many of the small new wetlands in northern Britain, however, have been carefully engineered. Their first major creators were the Victorians who built reservoirs of clean drinking water for the great industrial cities of Lancashire, Yorkshire, north-east England and central Scotland, as well as in Wales, once Edwin Chadwick and others had convinced society of the absolute need for abundant clean water and the scandal of its absence. Everyone knows the names of the great canal, bridge, road and rail builders of the Industrial Revolution, from John Loudon McAdam and Thomas Telford to Isambard Kingdom Brunel and Robert Stephenson, but who could place J. F. La Trobe Bateman, Thomas

A river saved: the Spey at the Insh marshes SSSI, 1987.
Peter Wakely, English Nature.

Hawksley, James Leslie or George Leather, though their achievements in constructing a network for the storage and transport of water over the length and breadth of Britain are surely on the same scale? These four men and their firms alone were responsible for eighty-eight constructions big enough to be included in the International Register of Large Dams, the Bateman office itself building forty-three of them.[32] The plunging death-rate from typhoid and cholera was their monument as surely as the soaring dams they built.

Little towns built little pools. Big towns built astonishing feats of engineering. At one extreme were the small and middle-sized burghs of Fife, ranging from Dunfermline, Kirkcaldy and St Andrews down to the diminutive Anstruther Wester with fewer than a thousand inhabitants: when the rage for a gravitational water supply struck, each built a separate reservoir within ten miles of the town, even Anstruther Wester refusing initially to join with contiguous Anstruther Easter and Cellardyke.[33] As late as 1968, Fife and Kinross was still being served by eighteen different water authorities before being swept up into one Water Board.[34] Those reservoirs that are not now decommissioned provide a dozen or so small and middle-sized wetlands in an otherwise dry agricultural countryside, and redress a little the balance of history.

At the other extreme are the mighty works of Manchester, Liverpool and Glasgow, taking water from the remote, unpolluted uplands of the Lake District, Lake Vyrnwy in North Wales and the Trossachs respectively, and conducting them by tunnel and aqueduct up to a hundred miles, in the case of Manchester, to the point of consumption. Such great undertakings were fraught with problems, both technical and political. On the technical side there was, to start with, an alarming degree of trial and error. Even the greatest of conventional engineers had little clue about the flow of water. Brunel, for example, advised that 'the form of sewers has practically very little to do with the general question of their keeping clear of deposit', when the opposite was the case, and Robert Stephenson opposed what turned out to be the highly successful glazed self-cleaning pipes: 'as to pipes he would not touch one; he hated the very name of them and felt inclined never to mention the word again'.[35]

Worse, very little was initially known about large dam construction. There were two serious disasters, when George Leather's Bilberry Dam near Huddersfield broke in 1852, killing eighty-one people, and when his

A river spoiled: the Irwell at Bury, Lancashire, 1989, with foam
effluent from a paper factory upstream.
Peter Wakely, English Nature.

nephew, John Towlerton Leather's Dale Dyke dam broke near Sheffield in 1864, drowning nearly 250 people. La Trobe Bateman in 1852 narrowly averted what would have been an even bigger disaster when the half-built Woodhead and Torside dams fifteen miles above Manchester were swamped with a flash flood. Both Bilberry dam and Dale Dyke dam disasters caused more deaths than the collapse of the Tay Rail Bridge, but it is somehow characteristic of the history of water engineering that the former have been totally forgotten and the latter vividly remembered. Even after Dale Dyke, Bateman, by then the leading reservoir engineer in Britain, continued to use a basically unsafe design of discharge culvert through the dam wall, though his standards of workmanship in engineering execution were so high that none presented serious problems in his lifetime.[36]

The political problems encountered by the engineers were exemplified by the opposition to the scheme promoted by Glasgow Corporation to bring the waters of Loch Katrine over thirty-four miles into the city (see Figure 4.3). When it was mooted in 1852, opposition was stirred up by the existing local water companies involved in pumping water to supply the city from the River Clyde. First, they persuaded the Forth Commissioners, and through them the Admiralty, that taking water out of one river basin into another would bring to an end the scouring action of the Forth and ultimately silt up the Royal Navy base at Rosyth: this fantastic suggestion perhaps underlines how little was known about the science of water flow. Despite an independent assessment from Brunel and Stephenson (more accurate this time) that such silting was impossible, it needed a visit from the Lord Provost of Glasgow to Lord Palmerston before the Admiralty withdrew its objection, and a sweetener of £7,000 to the Forth Commissioners before they would withdraw theirs. More serious still was the allegation by the water companies supported by the distinguished local chemist Professor Penney that soft mountain water would dissolve lead in the pipes that connected the houses with the mains.[37] This was refuted by the complex arguments of other chemists retained by the Corporation. That Penney was correct became only too apparent when lead poisoning was identified in Glasgow children about a century later, but the immediately dwindling death-rate from water-borne diseases in the city indicated that the gain to health by replacing Clyde impurity with Trossachs purity was both much greater and more immediate than the final damage caused by the lead.

Bringing the water of Loch Katrine to Glasgow was an operation conducted with characteristic Victorian panache. The surface of the loch was raised by four feet and water drawn off three feet below the natural water level, so that the storage in the loch was an upper surface seven feet deep. Compensation was provided for the Forth by a further reservoir downstream

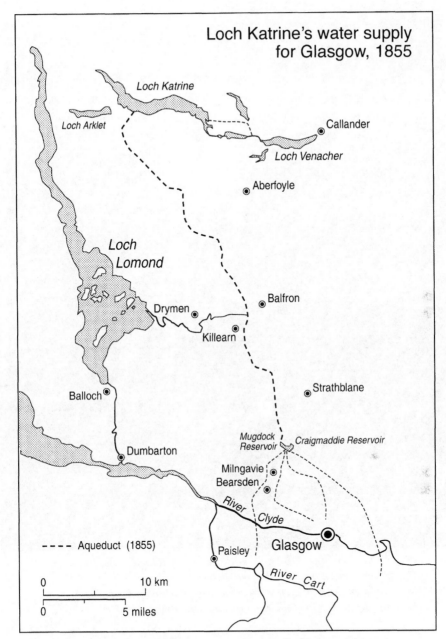

Figure 4.3 Loch Katrine's water supply for Glasgow, 1855.

at Loch Venachar. A tunnel was then driven through the ridge between Loch Chon and Loch Katrine to bring the water into a different catchment, and a temporary village was built in the very remote area at the head of

*Thirlmere. In 1876 La Trobe Bateman first proposed flooding
a valley in the English Lake District to supply Manchester:
'It could not have been more exquisitely designed for the purpose.'
Cambridge University Collaction of Air Photographs. Copyright reserved.*

Loch Chon, housing 3,000 employees and their families, with provision
stores, reading rooms, schoolhouse, church, resident doctor and teacher:
the labourers called it Sebastopol from the incessant roar of gunpowder
used in the tunnelling. Nearly four miles of the route was iron piping, the
remainder arched aqueducts: there were seventy separate tunnels, forty-four

ventilation shafts and twenty-six significant iron and masonry aqueducts.[38] Bateman said at its opening:

> I leave you a work . . . as indestructible as the hills through which it has been carried – a truly Roman work . . . It is a work which surpasses the greatest of the nine famous aqueducts which fed the city of Rome, and amongst the works of ornament or usefulness for which your city is now distinguished . . . none . . . will be counted more creditable to your wisdom, more worthy of your liberality or more beneficial in its results than the Loch Katrine water works.[39]

All went swimmingly until the taps were turned on in the Glasgow houses. In 1860, James Gale, the Corporation's chief resident water engineer, reported that on average thirty gallons of water per individual per day were leaking away due to faulty taps and cocks, 'considerably more than a half of the 50 gallons a day, which is now the average supply to each inhabitant on the north side of the Clyde'. Four years later he reported revised figures of 15 gallons wasted and 42½ supplied. To put this in perspective, Glasgow before the Loch Katrine scheme already took about 36 gallons per head (albeit of dirty Clyde water); Manchester in 1864 was using 22 gallons per head, and Sunderland 15 gallons per head, or as much as leaked away from the Glasgow taps even then.[40] Water waste was obviously a long-standing Glasgow habit. In other places, however, it was not so much the purity of water as its availability that mattered most in ordinary life. Gateshead in 1845 was African in its problems: a majority of the population was accustomed to wait up to three hours for a bucket at the public well. The spread of abundant domestic piped water has been called by Caroline Davidson 'undoubtedly the most far reaching change in housework in Britain between 1650 and 1950'.[41]

By 1895, a new aqueduct had been built from Loch Katrine to Glasgow and the loch level raised a further five feet, partly because Bateman's design produced less than promised, partly due to expanded industry and population. Domestic consumption per head that year, despite the great increase in baths, fountains, urinals and water closets in the city, had actually dropped to about 32 gallons a day as water fittings were brought up to standard.[42] That is close to (but higher than) the modern figure for domestic water consumption per head in Britain – 29 gallons in 1985.[43]

New reservoirs and water extraction schemes continued to be built in Scotland and the north of England in the twentieth century, though there have been few in the last thirty years. There were 380 drinking-water reservoirs and lochs in commission in 1971 in Scotland alone. It is, however, to the construction of the Scottish hydro-electricity schemes that we must look for the most remarkable feats of the water engineers in the twentieth century. The first industrial scheme of this kind was completed for the British Aluminium Company at Foyers in 1896, immediately followed and eclipsed by their scheme at Kinlochleven in 1909, involving water impoundment on an unprecedented scale in the British Isles – the Blackwater Dam was 86 feet high, half a mile long and held back 24 million gallons: it was, it has been said, 'the last major creation of the traditional navvy', and the beastly conditions under which they worked were described in the autobiographical novel of one of them, Patrick MacGill's *Children of the Dead End*.[44]

During the interwar years there were a number of other private schemes, notably in Lochaber, Loch Ericht, on the River Tummel and in Galloway, but it was left to the energies of the North of Scotland Hydro-Electric Board, set up by government in 1943, to transform the character of natural waterflow in the Highlands. Over the next twenty years they built fifty-three dams and power stations and installed a generating capacity of over a million kilowatts.[45] By the time they had finished, there was scarcely one really large natural water body left in Highland Scotland, apart from Loch Maree, which was untouched by water-impoundment or water-extraction schemes, and hardly a Highland river whose flow was not affected. Rainfall naturally bound for the North Sea was readily diverted from one catchment to another through tunnels and turbines into the Atlantic. It was all on a scale that showed La Trobe Bateman to have been a mere tyro in the art of commanding the waters.

Now, the usefulness of the skills of the water engineer is hardly in question, though it was hard-nosed cost-benefit analysis which in the 1960s in Scotland finally curbed the exuberance of the Hydro Board, and in England brought the Water Resources Board to an end after it had gone on an 'engineering spree' and sponsored the huge Kielder Reservoir to supply a steel works in the north-east that never got built.[46] But water is also delightful – to the angler, the naturalist and the rambler. To the romantic who may be one or all of these, looking with the eye trained by Wordsworth or Scott, some water composes classic landscape. What happened when use and delight came into conflict?

The answer could be expected to differ depending on the water concerned. River pollution triggers a quite different and wider response than reservoir construction, primarily because public health and utility is damaged by the first but furthered by the second. Protests against river pollution became commonplace in the mid-nineteenth century, and increasingly anguished, exemplified by the angry Yorkshire manufacturer who wrote to government using the water of the River Calder as ink: 'Could the odour only accompany this sheet, it would add much to the

An outing for the Corporation, 1880: the forty-two members of the
Glasgow Water Commissioners visiting a waterworks.
West of Scotland Water.

interest of this memorandum.'[47] Nevertheless, the convenience to industries and municipalities in having a river in which to dump waste quickly
outweighed complaints, in the judgement of local officials, especially when
most towns stopped drinking river water in favour of upland reservoir
water; they then felt free to pour greatly increased quantities of foul water
into the rivers without giving the consequences much thought. Even the
richest and most powerful private individuals in the land found it almost
impossible to get redress. It took the Duke of Buccleuch a generation to get
a judgement against the paper-makers upstream of his estate on the River
Esk in Midlothian, and in Yorkshire the law was so ineffective that the
Royal Commission on River Pollution in 1867 said that a person examining
the stinking rivers of the West Riding, 'would conclude that there existed
a general license to commit every kind of river abuse'.[48] Legislation from
1875 and 1876 in England (1867 in Scotland) recognised that there was a
serious problem but unsurprisingly failed to solve it at a stroke.[49]

Consequently, though much was done in the interim, the middle decades
of the twentieth century still saw scenes reminiscent of the nineteenth.
Kempster could write in 1948 that 'such industrialised rivers as the Tyne,
Tees and Wear' were 'in certain parts little more than open sewers'. He
noted that the Tyne at Newcastle, for several days in June 1933, had been
found to contain no oxygen at all, and that 'the crude sewage of several

towns goes untreated into the river'. Scotland was on the whole better, but he observed that there was 'little excuse for the pollution of such fine Highland streams as, say, the Deveron, into which the towns of Huntley and Keith discharge their crude sewage'.[50] One problem was the very large number of drainage authorities with conflicting objectives and inadequate powers: in the interwar years there were, for example, one hundred authorities within a fifteen-mile radius of Manchester Town Hall.[51] It is therefore not surprising that rapid deterioration in the face of industrial development was still the norm. Into the Tees, once an exceptional salmon river, Darlington in the 1940s still poured three million gallons of sewage a day, in addition to it receiving the effluent from coke ovens higher up and chemical works lower down. The estuary was poisoned with cyanide, and migrating fish sold at Hartlepool dropped from 180,400 lb to 700 lb between 1923 and 1937.[52]

Where it was feasible, cities disposed of their sewage straight into the sea: the estuary of the Mersey received 40 million gallons of raw sewage a day in 1935, and Edinburgh continued to pour raw sewage into the Firth of Forth from short outfalls until the early 1980s. When it changed the method of disposal, populations of species of polychaete worms characteristic of polluted waters were replaced by others characteristic of clean beaches.[53] But one of the famous sights of the Forth for bird-watchers, flocks of up to 10,000 scaup and almost as many pochard, disappeared overnight. The ducks had fed on the first population of worms and had no use for the second. Such are the ironies of conservation science.

Since 1948, but especially in 1951 and 1974, a series of new statutes have helped to reduce river pollution, despite new problems like discharge from intensified animal husbandry, pesticide and fertiliser run-off from arable farming and old ones like overloaded municipal sewage works, a problem which, for example, dogged the West Riding Rivers Board from its inception in the 1890s and dogs its successor Yorkshire Water today.[54] In England, improvement has been clearest, firstly on grossly polluted rivers (defined as those unable to support fish) and secondly on tidal rivers, worse than non-tidal ones before 1970 but distinctly better since. Scotland in 1985 had 95 per cent of river mileage classified as good, compared to 69 per cent in England and Wales. This was the best in the then European Community, but reflects the non-industrial history of all but the central belt rather than some prime Scottish virtue of keeping the water clean.[55] Nevertheless, Britain remains better at dealing with spot pollution, as from a city or a factory, than with diffuse pollution, as from agricultural run-off. The River Ythan in rural Aberdeenshire was gravely polluted by farm nitrogen in the late 1980s, and Loch Leven in Kinross lost its famous brown trout fishery around then, for similar reasons. In so far as there has been a breakthrough

*The achievement of the engineers: the great puddle trench at
Craigmaddie Reservoir was 193 feet deep and an integral part of
bringing Loch Katrine to Glasgow. The figure in the top hat is probably
J. M. Gale, chief engineer to the Water Commissioners.
West of Scotland Water.*

in law and effective action, it came after the Second World War and especially in the 1970s. Considering the antiquity and potential strength of the
anti-pollution lobby, it seems at first sight quite surprising that delight
made such slow progress against use.

When one similarly considers criticism of large-scale water impoundment,
there is little to report before the end of the nineteenth century. Walter

Scott's epic poem *Lady of the Lake* had from 1810 onwards made Loch Katrine the most sacred spot on the Scottish romantic tour – when statistical Sir John Sinclair went there in the autumn of that year he found his carriage to be the 297th of the season; by mid-century the loch had thousands of visitors and steamboats plying the water.[56] Yet no one raised an audible protest about its impounding by Bateman which began in 1856 or its further raising by the Act of 1885, though the pumping-stations and dams were given suitable baronial and rustic trimmings to disarm criticism. Neighbouring Glen Finglas, once almost equally famous, likewise began to be impounded without comment from the public as late as 1958.

It was the Lake District, perhaps predictably given the potency of Wordsworth's ghost, that sparked the first row. In 1876, La Trobe Bateman proposed tapping Thirlmere and bringing its waters one hundred miles to Manchester. The corporation, having previously sent up three spies disguised as a cattle dealer, a property agent and a tourist writer, decided to purchase the entire catchment in order to keep the water pure. The Bishop of Manchester, expressing a view at odds with that of popular theology in the Lakes, held that if Thirlmere 'had been made by the Almighty expressly to supply the densely populated district of Manchester with pure water, it could not have been more exquisitely designed for the purpose'. Local owners and inhabitants were outraged: the Thirlmere Defence Association was formed objecting to large-scale engineering in the Lake District, and thirty-three petitions were heard by a specially constituted Select Committee in Parliament. It was the first time such a committee had examined not only property interests but impact on landscape and enjoyment, foreshadowing modern public inquiries. The proposal was approved in 1879, but due to trade conditions work did not start until 1890 and the dam was not completed until 1894.[57] In the same year, appalled at the reality of the damage, Canon Rawnsley and his allies in Lake District defence, including the redoubtable and energetic Octavia Hill, had formed the 'National Trust for Historic Sites and Natural Scenery' to act as custodian for property acquired for the nation. A large part of the agenda was to save the Lakes from further exploitation of this kind. It was nevertheless unable to prevent the sale of Haweswater and its catchment to Manchester in 1916, though water did not flow from there until 1941 and much of the controversy in the Lakes in the interwar years was directed at afforestation rather than reservoirs.[58]

After the Second World War, Manchester Corporation engineers, full of hubris, and conscious of their undefeated record in the Lakes, turned their attention to Ullswater. In 1961 their attempt to get a private bill to allow them to raise the water failed in the House of Lords due to the bitter and skilful opposition of the experienced judge and counsel, Lord Birkett, as the spokesman for the objectors, and was not resurrected. In 1964, Manchester tried again through a water order to get permission to abstract from both Ullswater and Windermere. The ensuing public inquiry was heard in Kendal, where local people refused to house or feed the proponents of the scheme. Permission to use the lakes was ultimately granted by the minister, but under conditions of unparalleled stringency: the pumping stations had to be underground, the weir at Ullswater was not allowed and proposals for a reservoir in Barrisdale were abandoned. All this added greatly to the cost. Delight had shown teeth.[59]

Opposition to hydro-electric dams took a somewhat similar course. The 1896 Foyers scheme interfered with a famous landscape and a waterfall that had inspired Burns, but it was warmly endorsed at a public meeting, 'two ladies alone dissenting', as the 'impoverished crofters and fisher folk of the Western Highlands' acclaimed 'the advent of this industry into their midst'. In the interwar period the amenity lobby grew, and in combination with salmon fishing and coal-mining interests, managed to block five bills to develop the Highlands between 1929 and 1941. Much of the support for amenity was based in England, but the Association for the Preservation of Rural Scotland in 1936 demanded an inquiry into whether the alleged benefits of hydro schemes would really outweigh the injury inflicted on the countryside.[60] Feelings ran high, even in wartime. The debate in the House of Commons on the Grampian Electricity Supply Order in 1941 was impassioned. Noel Baker, a powerful English Labour MP and future minister, spoke of proposals to flood Glen Affric and Glen Cannich as:

not merely a matter for the people of the Highlands. This is a matter for the whole British nation . . . not a question of minor or transient aims. The Highlands are the spiritual heritage of the whole people which it is the duty of Parliament to preserve.[61]

Tom Johnston, Secretary of State for Scotland, after this debate commissioned the Cooper Report to examine the future for hydro power in the Highlands. It did not pussyfoot over the matter of delight versus use:

The final issue in this matter must be faced once for all realistically. If it is desired to preserve the natural features of the Highlands unchanged in all time coming for the benefit of those holiday-makers who wish to contemplate them in their natural state during the comparatively brief season imposed by the climatic conditions, then the logical outcome of such an aesthetic policy would be to convert the greater part of the area into a national park and to sterilise it in perpetuity, providing a few 'reservations' in which the dwindling remnants of the native population could for a time continue to reside until they eventually became extinct.[62]

The Cooper Report was followed by the establishment of the Hydro Board in 1943, but consciousness of the continuing strength of feeling persuaded the new body to go to great lengths to try to pacify sporting interests by the liberal provision of fish ladders, and amenity interests by carefully designed dams and power stations, some of which (as in Glen Affric) rank among the better architecture of the postwar period.

The Board could not, however, avoid a full-scale battle in 1945 with the National Trust for Scotland and others over its proposals to create a new reservoir – Loch Faskally – at Pitlochry, another occasion where proponents were refused hotel room and food by the local community. The ensuing public inquiry found in favour of the proposal, and the NTS, saddled with large legal expenses, went away to lick its wounds for more than a decade. Glen Cannich and Glen Affric were flooded without formal objection from them, and dams appeared the length and breadth of the Highlands. In 1958, Glen Nevis began to be considered for a hydro scheme, and the Trust girded its loins for another fight: 'there is no compromise over this', wrote its secretary, stiffening the chairman to save 'the attractive hanging valley immediately above the gorge and a 400-foot waterfall at the head of the valley'.[63] They prepared to fight at a public inquiry, but in the event did not need to, and instead gave evidence to the Mackenzie Committee appointed in 1961 to inquire into the future of electricity generation in Scotland. The trust, in an important submission, revived an old argument from the National Parks debate in Scotland that there was a need for a Landscape Commission with powers to designate areas worthy of special protection – a foreshadowing of the Countryside Commission for Scotland.[64] The Glen Nevis proposal was eventually withdrawn in 1965, scuppered on economic grounds in a changed climate of assessment for hydro projects, but it could not but be seen also as a victory for the amenity lobby.

By the mid-1960s, then, it was for the first time no longer self-evident that impoundment, generation and extraction schemes would any longer get their way. Following the most anguished debate, a proposal to flood sugar limestone of the highest botanical importance at Cow Green in Teesdale was pushed through Parliament in 1967, the need to service ICI's projected new ammonia works outweighing the united opposition of the Nature Conservancy, the British Ecological Society and a wide range of voluntary bodies. As John Sheail remarks, for the objectors 'it was a poor consolation to know that things could never be quite the same again', and that projects like this in the future would have to take ecological considerations much more seriously if they were to avoid expensive argument and adverse publicity.[65]

It was a turning point. In Yorkshire, proposals in 1969 for water impoundment in the Hebden valley, very close to a site rejected twenty years earlier on amenity grounds, was defeated in the House of Commons after Douglas Houghton, the local Labour MP, pointed out that it would be the eighteenth reservoir in his constituency and called on Parliament to 'take a stand on natural beauty in an area of ugly legacies'.[66] A scheme in the following year to build a reservoir at Farndale in the North York Moors National Park was dramatically turned down.[67] In the Peak District, which supplied four million people from 'a lakeland but an artificial one – a distinctive water-supply landscape', it had been argued in 1962 that 'most people would agree' that the reservoirs had improved the scenery.[68] This was reminiscent of the Hydro Board's claim of public recognition that flooding Glen Affric had increased its beauty by 'some hundred per cent'.[69] But proposals ten years later to build another huge impoundment in the Peak National Park resulted in three expensive public inquiries before, in 1978, the scheme at Carsington outside the park boundaries was finally approved.[70] Few large reservoirs have been built since. By the 1970s, to put it no higher, the proponents of delight had learned how thoroughly to gum up those schemes of use which they deemed unacceptable. Though outside our chosen area, a particularly interesting case was Rutland Water, made acceptable in 1969 by statutory provision to construct on site not only generous provision for recreation but also a large nature reserve to add to the biodiversity of the area – a provision that has been strikingly successful.

A final observation is in order. That river pollution, which has much to do with environmental and therefore public health, should only be effectively tackled at much the same time as unacceptable water impoundment, which is primarily to do with aesthetics (and very occasionally nature conservation), strikes me as remarkable. It argues that there was, around the 1960s and 1970s, a real paradigm shift in our constructions, or perceptions, of nature, enough to move politicians and institutions, not just a change in fashion or sectoral influence empowering one narrow lobby or another for a moment. But that is a matter to return to later.

CHAPTER 5

THE FRAGILE HILL

Most people, if asked what they considered the wildest part of Great Britain, would probably name the mountains and moors which make up over half the land surface of northern England and Scotland. If wildness means the chance of walking for hours without encountering a road or any habitation other than the occasional farm, and that probably now deserted, this is correct. Yet not an inch of the uplands is unmarked by human imprint. The most natural areas, those where we can enjoy a view closest to the experience of the first peoples in the post-glacial wilderness, are the mountain tops, the hill lochans and tarns, and the great blanket mires – yet they also contain, as we shall see, an embedded archive of our activities as a polluting species.

The least natural areas of the uplands are the sweeping, open stretches of moor, which generally came about through human action and would mostly disappear without its continuation. Characteristically the moors originated millennia ago. In the Lake District, for example, archaeological exploration has shown how, in the Bronze Age, the twin features of a wetter, windier climate and pressures from domestic animals led to the failure of regeneration in the primal oakwoods, and eventually to the replacement of the trees by a pasture of fine grass. But then, as the ground, now unprotected by a canopy and unfertilised by leaf fall, became progressively more leached of nutrients, more acidified and podsolised, the grass was replaced by heather, bilberry and other ericaceous shrubs.[1] Such heather (*Calluna*) moors have been maintained since their creation by the tooth of sheep, cattle, goat, pony and deer, and by the custom of burning at intervals to stimulate new growth. Were animals and fire to be withdrawn, as might well happen in the face of prolonged agricultural depression, tree cover in most places would return naturally. It would not now be oak, but species tolerant of the acidified soil – birch, Scots pine and various species of exotic conifer such as Sitka spruce and lodgepole pine spreading from adjacent plantations.

Heather has in any case been in rapid decline in the last fifty years over most of northern Europe, not because of such losses to natural vegetation succession but because of further human intervention: new conifer forests planted on the moor, losses to improved fertilised pasture at the farm edge, overgrazing by sheep and the effects of acid rain. Scotland lost about a quarter of its heather moor and heather-dominated blanket mire between the 1940s and the 1980s. We shall return to some of these causes of this loss below. And not all moorland was dominated by heather to start with. The wettest areas of the uplands, those in the west with heaviest rainfall or those on impermeable soils, became dominated, once the original tree cover had gone either due to human action, climatic change or a combination of both, by mat grass (*Nardus*), purple moor grass (*Molinia*), cotton sedge (*Eriophorum*) or sphagnum moss. Here future reversion to tree cover would not be so natural or so speedy were human intervention to cease, because a thick entanglement of coarse grass or a wet bog do not form a comfortable seedbed for regenerating trees.

For the last two centuries, for whatever reason, coarse grasses and sedge have been spreading on the northern British uplands at the expense of heather and sphagnum. In the southern Pennines, for example, in 1813 a commentator spoke of the Derbyshire moors covered by 'great accumulation of the grey Bog-moss, *Sphagnum palustre*', and in 1835 another naturalist described *sphagnum's* ecological associate, the beautiful bog rosemary, as 'abundant in all that group of mountains that separates Yorkshire from Lancashire'.[2] *Sphagnum* and bog rosemary are today absent from all that area where they once grew so freely, replaced by cotton sedge. But such constant change is, or has become, a feature of moors.

This century there has been much scientific discussion about the 'degradation' of the uplands, most memorably when Fraser Darling linked deforestation and over-exploitation to notions of desertification popular in the wake of the American dustbowl experiences of the interwar years. In 1956 he wrote that in Alaska he had hit upon the germ of an important idea that:

> The course of overgrazing and overburning which has occurred over so much of the world has produced vast areas of desert: I myself have demon- strated that a wet desert has been produced by this means in the Scottish Highlands.[3]

An ecologist might take this view, observing how a wooded environment rich in nutrients and in a 'long-continued steady state', can become trans- formed by primary deforestation into something else. Pearsall, for example, explained how in the Lake District the accumulation over millennia of leaf mould and humus in the soil in a short time became oxidised and destroyed, forming soluble salts that combined with the available lime and potash in the soil itself and ended up in the lake sediments, leaving the hillsides acid and bereft of nutrients.[4]

Yet all this generally happened long ago, longer ago than Fraser Darling

Nature Conservancy Warden Polson scanning Beinn Eighe
National Nature Reserve, Wester Ross, for deer, 1950s.
English Nature.

thought. Birks, for example, summarises the fossil pollen evidence for the earliest extensive clearance: in the north-west Highlands 3,700 years ago; in Skye between 3,900 and 1,700 years ago; in the North York Moors and the Pennines 2,500 to 2,300 years ago; in the Lake District, Galloway and Argyll 1,700 to 1,400 years ago.[5] Only clearance in the Grampians and the Cairngorms is dated by him to as recently as three or four hundred years before the present, but there is other palynological evidence to suggest that in many localities even here it may go back much earlier, into prehistory.[6]

That being the case, the transformed uplands have had a long time to establish their own distinctive ecosystems adapted to a nutrient-poor environment, and these, too, are highly valued. One quarter of all the British invertebrates, for example, are found on heather moors. There are at least sixty-seven species of upland breeding birds, a quarter of which can be considered rare or at risk, and sixty-one are associated either with *Calluna* heaths, grassy moors or acid bogs.[7] There is probably a greater mixture of bird species typical of different climatic zones breeding in the British uplands than in any other comparably sized part of Europe.[8] But there has also been a great deal of anxiety recently about moorland and montane habitats themselves becoming biologically less productive, that is capable of supporting less life, and that of fewer species. It is in this sense, and not in the sense of the primary conversion from woodland to moor, that what happened in the last four hundred years is considered here.

Evidence for the decline of the biological productivity of the hills is often impressionistic but compelling. The history of birds (better known than that of other groups) provides some examples. Wheatears are characteristic upland birds, though in the Highlands today you could easily walk for an hour in summer without encountering a pair. They migrate south. In the eighteenth century Gilbert White reported that on the South Downs in Sussex:

> They so abound in the autumn as to be a considerable perquisite to the shepherds that take them . . . at the time of the wheat-harvest they begin to be taken in great numbers . . . and appear at the tables of all the gentry that entertain with any degree of elegance . . . in the height of the season so many hundreds of dozens are taken.[9]

Others reported that a shepherd had snared a thousand in one day, and that about 22,000 were taken annually around Eastbourne.[10] After 1850,

Two-year-old red deer stag scanning Inverinate Forest,
Wester Ross, for cyclists, 1934.
St Andrews University Library, Adam Collection.

however, many shepherds ceased setting traps as the yield was not worth the trouble. It is hard to imagine that bird ringers on the Downs today, benevolent modern equivalents of those predatory shepherds, would encounter more than a trickle of birds in a season. Very many of these wheatears would have been migrants from the north, though others would no doubt have come from southern heaths like the East Anglian Brecks where the decline has been even sharper than in the hills.

Gilbert White was similarly the first to describe the regular migration through the south of another northern British breeding bird, the ring ousel. He reported flocks in Hampshire of twenty or thirty birds in autumn, and had two shot in spring. He ate one and found it 'juicy and well-flavoured'. The ring ousel is a rarity in the area today (just four were reported inland in autumn in the 1996 *Hampshire Bird Report*). It was then also much commoner on its nesting range. White reported that 'they breed in great abundance all over the Peak of Derby and are called there Tor Ousels'. On the North York Moors, the elderly rector of Danby in 1907 recalled formerly having encountered flocks of 'some hundreds' of ring ousels in September, and of the rectory garden regularly being invaded by fifty at a time once they had finished the bilberries on the moor; by 1969 the species' status in the area could only be described as 'very local'. In Scotland, James Robertson in 1771 found them in 'great numbers' in 'the vallies among the snowy mountains' in upper Deeside in June , feeding on juniper and 'most numerous where these berries were most abundant'.[11]

They are certainly not omnipresent in the same way in such places in Scotland or in northern England today. One historian of bird distribution considers that numbers of ring ousel remained 'stable during the nineteenth century. . . [but] started a long and steady decline early in the twentieth century'.[12] This has been attributed partly to disturbance by ramblers, but there were also high levels of human disturbance in the eighteenth century from shepherds and others at the summer shielings. It seems much more likely that some deterioration of the habitat lies at the root of their decline.

The dotterel is a third northern species, this time of the high tops, the decline of which certainly predates the twentieth century. It also was once so common on migration as to attract the attention of hunters and gourmets. 'Great men and kings are keen in the chase of this bird,' said Isaac Casaubon in 1611; 'it furnishes very delicate meat, if my palate is sufficiently instructed.'[13] In the eighteenth and earlier nineteenth centuries, flocks of hundreds were recorded on spring migration from Aberdeenshire to Cambridgeshire. In southern Scotland, 'gunners went out and slaughtered them in numbers: they were very tame on first arrival'.[14] In Yorkshire many hundreds were still shot around 1850 on the spring passage on the Holderness coast, and on the wolds, moors and commons inland. The Dotterel Inn at Reighton was purpose-built by the Strickland family 'for the accommodation of gamekeepers who came from all parts to the neighbourhood' to shoot them in spring. Their feathers were much sought after by Victorian fishermen, and the species began to decline steeply in the middle decades of the nineteenth century.[15]

From then on they became scarcer and scarcer. Fewer than a thousand pairs of dotterel now breed in northern Britain (almost all in Scotland), despite some recent increase. A flock of a hundred has probably not been seen in Britain since the First World War, and by then such numbers were very exceptional. To see a dotterel on migration now is something of a prize to a twitcher, and the 'trips', or little flocks, are invariably small.

Many descriptions of the uplands before about 1870 suggest a general profusion of life at odds with experience today. Thus Thomas Pennant, travelling in upper Deeside, in the 1760s described the zone above the pine woods in terms that recall the first American travellers to the west:

> The whole tract abounds with game: the stags at this time were ranging in the mountains, but the little roebucks were perpetually bounding before us; and the black game often sprung under our feet. The tops of the hills swarmed with grous and ptarmigans. Green plovers, whimbrels and snowflecks breed here . . . [16]

– and so on. A century later, on the other side of Scotland, in Gairloch, Osgood Mackenzie quoted his game book for 1868:

> My total for that year was 1,314 grouse, 33 blackgame, 49 partridges, 110 golden plover, 35 wild ducks, 53 snipe, 91 rock-pigeons, 184 hares, without mentioning geese, teal, ptarmigan and roe etc., a total of 1,900 head. In other seasons I got sometimes as many as 96 partridges, 106 snipe and 95 woodcock. Now [1921] so many of these good beasts and birds are either quite extinct or on the very verge of becoming so.[17]

At first sight it seems a clear case of cause and effect, but it was not obvious to Mackenzie that it was his own shooting that caused the decline, and we should not take it for granted either.

Game bags are important because they can be quantified. The most carefully worked out statistics for declining moorland productivity for any species come from Peter Hudson's studies of red grouse, 1890–1990. Over this period the mean bag per square kilometre fell by about 40 per cent across northern Britain, though not evenly in time or in space. The fall was steepest in the 1910s, between 1935 and 1945 and in the 1970s, with limited recovery between. It was much more evident in Scotland than in northern England. In the east of Scotland, for example, there was a fall of over 70 per cent between around 1910 and around 1985, in the west of Scotland (usually on less productive moors) a fall of about 90 per cent, but in the North Yorkshire Dales and in Cumbria bags were actually higher in the 1970s than at any earlier period this century. The English moors have always been more productive of grouse than the Scottish moors, and the south and east generally more so than the north and west.[18]

The red grouse has been described as a 'test animal', so that the decline in its numbers, in McVean and Lockie's words, 'can be regarded as some indication of the extent of habitat destruction which has taken place over the last century or so'.[19] This may be broadly true, but all sorts of factors other than habitat degradation can be responsible for short- and medium-term variation – the number of keepers, the number of predators (clearly related), effort put into shooting and management, cycles of disease and so forth. The spatial as well as the temporal complexity of the pattern warns us against monocausal explanation either for the decline of red grouse bags or for any other symptom of declining upland productivity. It is also important to appreciate that Hudson's study only starts in 1890 and for many areas there are little data until after the First World War. Prior to that date, or at least in the decades after 1860, bags had increased steeply as driving replaced shooting over dogs, and muirburn to improve the habitat for grouse became more systematic.

It is therefore particularly useful that there has survived a very remarkable series of gamebooks from the extensive Buccleuch estates in southern Scotland, which show the annual kill of game for every year since 1834 (even earlier in a few instances). Figure 5.1 gives the figures for red grouse shot at Drumlanrig and Sanquhar – rising from levels of fewer than 1,000 before 1850, to a dizzy and manifestly unsustainable peak of almost 15,000 on the eve of the First World War, and then dropping steadily back to the levels of the early nineteenth century by the 1980s. The figures from the Langholm and Newlands estates are similar in trend but even more extraordinary in steepness: before the late 1870s they amounted to a few hundred a year, and had fallen back to these levels by the late 1940s, but in the interim peaked at 29,000 in 1911.

The value of the Buccleuch red grouse statistics is enhanced by the other species recorded. Blackcock were a quarry for which the Buccleuch estates were especially famous – in some years in the nineteenth century they came close to equalling and occasionally even exceeded the tally of red grouse: the record shoot over all these estates was 4,300 in 1915. At

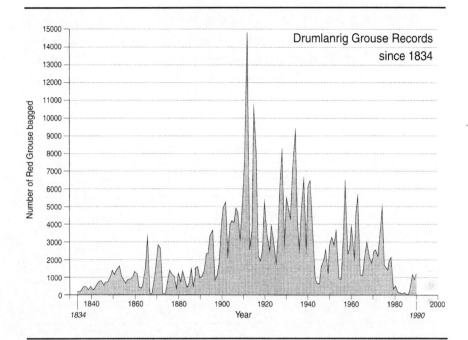

Figure 5.1 Drumlanrig grouse records since 1834.
Source: *Duke of Buccleuch's own calculations from the
Drumlanrig Game Books.*

Drumlanrig and Sanquhar they reached a peak around 1901 of 2,100, but fell like a stone in the 1930s to around a hundred. By the 1960s and 1970s they were half that, and by the 1990s they were not shot at all. In the whole of Britain probably only about 6,500 black grouse remain today. The prime cause of their decline is thought to be the destruction of bilberry and allied shrubs on which they depend, due to overgrazing by sheep.

Drumlanrig and Sanquhar encompass grounds that stretch from low straths to high moor, and some of the decline in game after the Second World War probably reflects what was happening in the valleys as much as in the uplands. Hares plunged from bags typically well over a thousand up to and including the 1950s, to fewer than 200 by the early 1990s, and partridges were down from several hundred to a couple of score (see Table 5.1). On the other hand, the pheasant, which lends itself to artificial rearing and helped to fill the economic gap left by the decline of wilder game, rose from a few hundred in the first half of the nineteenth century to well over 8,000 by the 1990s.

Another non-native species, the rabbit, which first arrived as a pest in these parts in the middle decades of the nineteenth century, multiplied very rapidly indeed: five was the average killed in 1834–8, but 87,000 were slain at Drumlanrig and Sanquhar in 1888 – admittedly a very exceptional total, but the annual average for that quinquennium was over 50,000. After years of very low figures following the arrival of myxomatosis in the 1950s numbers now seem to be returning towards their peak of a century ago.

Table 5.1 Average annual kill of selected game
at Drumlanrig and Sanquhar, 1834–1992

	Hares	Partridges	Pheasants
1834–8	1,175	372	283
1845–9	1,972	925	452
1859–63	3,537	505	1,144
1918–22	1,842	561	3,235
1930–4	2,633	599	5,125
1988–92	186	34	8,507

Source: Buccleuch muniments, Drumlanrig Castle.

The root causes of game decline must vary on different parts of the Buccleuch estates, patterns of modern arable farming possibly bearing much responsibility on the straths and overgrazing a likely cause on the tops. But the entire rural ecosystem of this part of Dumfriesshire is evidently much less productive for native wildlife than it was in the Victorian and Edwardian years.

To return to the broader question of the fortunes of the moors of northern Britain as a whole, one widespread phenomenon has been the protracted threat from changes in land use. Nearly 60 square miles were fenced and converted to agriculture or forestry in the North York Moors National Park itself between 1950 and 1975, and losses on this scale were repeated the length and breadth of the uplands. In Scotland, the area of heather moor fell by 23 per cent between the 1940s and the 1980s, most of the losses being to afforestation and grassland expansion.[20]

The central problem to be addressed here, however, is what happened to the productivity and character of the land that remained as moor. There have been three major influences since the mid-eighteenth century – the rise of the great sheep farms to replace a peasant cattle economy, the rise of the sporting estate and aerial pollution. Let us consider them in turn.

The great commercial sheep farm was nothing new in the late eighteenth century except in the Highlands. In Yorkshire and the Scottish Borders, large-scale monastic enterprise in wool production goes back to the thirteenth and fourteenth centuries, undoubtedly with concomitant changes in land use and local ecology. Lay lords and tenant farmers took up the example, already before the end of the Middle Ages producing far more wool than the church. After the Reformation production continued on a very large scale, now more for the burgeoning home textile industries than for sale abroad. This is worth emphasis, since it implies that the introduction of sheep cannot in itself be responsible for degradation and change over the entire northern British uplands in the last two and a half centuries. But of course it could be responsible for degradation and change in the Scottish Highlands, or novel features introduced since 1750 to sheep management in northern Britain as a whole could be responsible for a wider degradation.

The story of the introduction of commercial sheep farming to the north, and of the associated Highland Clearances, has been told too often to need retelling here.[21] Suffice to say that it began around 1760 in Perthshire, engulfed Ross and Cromarty and Sutherland between 1793 and 1815, was further extended into the islands and other parts by 1850 (the last clearances of crofters took place around 1855) and began to lose impetus and prosperity by 1880, though sheep farming has remained an essential part of land use in the Highlands and throughout upland Britain during the twentieth century.

What did the rise of modern sheep farming since the end of the eighteenth century do to the hills? Critics have generally alleged two effects: first, that the sheep modified the vegetation to make it less diverse, and so capable of supporting less biodiversity of vertebrate and invertebrate life; second, that the sheep extracted much of the nutrient of the soil in their bone and meat, driven off to be consumed elsewhere, or else it was dissipated in the intensive muirburn associated with their management – what Fraser Darling called 'two centuries of extractive sheep farming' that 'reduced a rich resource to a state of desolation'.[22]

The second accusation is the more serious, because presumably less reversible. By reintroducing cattle, for instance, it might be possible to reverse modification of the vegetation, but no amount of revised grazing practices could quickly recover lost minerals. The kind of soil degradation postulated, however, has proved hard to confirm scientifically: attempts to calculate nutrient budgets on moorland have usually concluded that what is chemically lost in smoke and meat is more than replaced under modern conditions by chemicals deposited in rain, though with two important *caveats* – phosphorus (already rare on the uplands) is likely to decline as a result of heavy grazing, and the effect of grazing and burning on western Highland moors may be more deleterious than those on eastern moors.[23] Erosion from the hooves of too many animals greatly adds to the problem if there is over-stocking, as the modern headage subsidy encourages. It exposes the nutrients to leaching out in the rain and washing down to the rivers.[24]

It has to be said, moreover, that experiments in this area are difficult and not all conclusions agree. One group of scientists found at Moor House in the Pennines no sign of a difference in soil fertility between grazed and ungrazed plots, so no support for the hypothesis that sheep had reduced the fertility of the moors over a timescale of eleven to thirty years.[25] On the other hand, another group studied three upland grassland plots presumed to have been similar in 1859. On the field lightly grazed by sheep since then the total nitrogen in the top six inches of the soil was 70 per cent more than that on a field that had been heavily grazed and even lightly fertilised in the same period.[26] So the question is not yet closed.

Scientists are in no doubt, however, that sheep farming modifies the vegetation, or that it can in some circumstances result in a fall of productivity, for example by locking up nutrients below a mat of *Nardus* grass which itself is less nutritious than alternatives.[27] Sheep are very selective feeders. By choosing the shoots of young heather on acid ground, high densities of sheep facilitate the spread of *Nardus* and *Molinia* grasses. The effect is easy to observe at fence lines from Shetland to the Lake District: heather at the roadside which sheep cannot reach, and tussocky grass on the hill beyond where they graze. On English chalk or limestone and where there is lower density there is no such effect: on basic soils, greater floristic variety may result from grazing sheep destroying invasive thorn shrub or tall herbs. The action of their grazing, lawn-mower-like, cuts close to the root, again damaging heather more than grass. A cow, by contrast, bites higher and tears at the vegetation; a heavy animal, it punches holes in the turf where a sheep walks over the mat, and a cow's runny dung is more often absorbed by the ground, where a sheep's small, hard dung may be oxidised and its nutrients dispersed into the air.

For both animals, and, indeed, for deer and rabbits, the effect that they will have on vegetation depends on the density at which they occur: light grazing tends to favour heath and heavy grazing grassland.[28] Land hammered by overgrazing from too many herbivores has a quite different appearance from lightly grazed land – nothing survives except a few tough species that can endure being razed almost to the surface of the ground. Grazing was very much lighter in the eighteenth century than it became subsequently. James Robertson in 1771, travelling in Deeside, perfectly expressed the reason:

Skiddaw group SSSI, Cumbria, 1993: left of path, well-managed
grouse moor; right of path, vegetation overgrazed by sheep.
Peter Wakely, English Nature.

The small spots of arable land which are found in the vallies alone, bear a small proportion to the hilly and uncultivated parts on which the people depend for pasture . . . during the summer and autumn the pastures could maintain thrice the number [of animals], but they would perish during the winter or spring. Even the scanty stock to which the farmer confines himself is with difficulty preserved and not unfrequently some of them die for want of fodder. Before blaming too severely the indolence of this people . . . we must consider the difficulties under which they labour.[29]

The problem was solved by the introduction of turnips and later of artificial feedstuffs, either fed to the sheep on the Highland straths or fed to animals driven off to be wintered elsewhere. Because more breeding stock were kept over the winter, certainly after the 1830s, vastly more animals could range the hills in summer. The obliterating effects of so many more teeth were compounded by the wider use of fire to clear the land of old heather and encourage the sprouting of new. Certain types of habitat that were probably quite common in the eighteenth century, such as montane scrub of creeping willow, dwarf birch and juniper, as well as more rowan, holly and other birch, would have been particularly hard hit, and with them the ring ousel's supply of berries and insects.

There was one point in the nineteenth century when it seemed clear to many observers that the productivity of the hills in the northern Highlands was declining in relation to the number of sheep they could carry. James Macdonald, special reporter to *The Scotsman*, described the change in an account of Ross and Cromarty in 1877:

> Considerable portions of the grazings are becoming foggy [mossy] and rough and of little value as sheep pasture. We could point to one or two hirsels which carried stocks of from 1000 to 1100 over winter some twenty years ago and which will now winter scarcely 800.[30]

In the 1870s and 1880s comments of this kind became widespread and commonplace,[31] but the most careful observers noted that pasture deterioration was largely confined to the 'green places', that is to areas of former arable and in-bye where the evicted Highland peasants had spent (in some cases) centuries of effort in transferring nutrients from the hill to enhance fertility near the settlements. The sheep were now busy transferring the nutrients back again. As Macdonald put it in another article on Sutherland:

> The green land . . . was reclaimed and enriched by the many hundreds of small tenants who long occupied the straths in the interior of the county. About sixty years have elapsed since these tenants gave up their holdings to the fleecy tribe. During all that period the land thus left in good condition for raising green pasture has been constantly grazed by a heavy, hungry stock of sheep, that have browsed upon it all day, and spent the night on the higher and blacker land, where they have also left the richest of their droppings.[32]

Because of this, and because the way they grazed affected the vegetation, the net effect was to reduce the number of animals that the sheep-runs could carry under that particular system of management. But the loss of the 'green places' made by the past labour of a vanished people was a one-off effect that did not itself make sheep farming unviable. Between 1875 and 1966 government statistics showed that the density of sheep increased by over 50 per cent over much of northern Britain, including the northern Highlands; much of this was encouraged after 1945 by livestock subsidies paid per animal kept. Upland farmers were able to keep more sheep partly by converting land at the margins of the moor to seeded pasture, partly by more seasonal supplementary feeding, which increased the pressure on the moor itself. In other parts of the Highlands – for example, Perthshire and Argyll – there was, however, no change in the number of sheep kept over that century.[33]

From the Langholm game book, 1893–4. The caption reads:
'Ugh, they're all grey hens, I shan't shoot. N.B. roughly indicated according
to fact are 342 partridges, 41 cock pheasants, 54 grice and 2 old blackcock.'
Duke of Buccleuch.

In a recent summary, Alexander Mather has shown how the nineteenth-century belief that sheep were destroying the fertility of hill grazings persisted among agricultural scientists until around 1950, though with little hard evidence being produced. His own studies suggest that lambing performance, which partly depends on the nutritional level of the ewes, fell in places in the north of Scotland quite markedly between 1890 and

1980 – by one percentage point every four years in the west coast parish of Glenelg.[34] But there can be reasons for this other than the removal of chemicals from the ground – for instance the quality of shepherding and an inappropriate density of stock designed to catch the subsidy. One has to say that the intrinsic unsustainability of intensive sheep farming in such places is not proven, at least in the narrow sense that there is not yet widespread or incontrovertible evidence that irreversible damage has been done by it to the soil beneath the sward.

On the other hand, the impact on the flora of the uplands over the past two hundred years is beyond all question. Several scientific papers have quantified the long-term effect of high sheep densities on heather cover. One study in Co. Durham, for example, concluded that a summer grazing of three sheep per hectare was enough to suppress heather growth, but that most moors in the area by the 1970s had stocking well above this level.[35] Another in Co. Antrim found a rate above 1.0–1.6 sheep per hectare reduced heather cover and impoverished insect diversity on blanket bog. Densities in the Forest of Bowland in Lancashire, with important blanket mire, were far above this; on rough grazing there they rose from three to sixteen per hectare between 1957 and 1990. At Wasdale in the Lake District, heather began to undergo a 'real decline' after 1850, following an increase in sheep and a local change to all-the-year-round grazing as the animals were bred primarily for meat rather than wool.[36] Stevenson and Thompson have argued more generally from pollen studies in upland bogs that sheep farming has had a wide effect on heather moors traceable in different places back to the seventeenth or eighteenth centuries, and almost universal from 1850 to 1900. It proved to be sharpest in those areas that have had the longest tradition of intensive grazing, suggesting that the impact is cumulative.[37]

The observations of local shepherds and farmers often agree with the scientists. The experienced sheep farmer Reay Clarke contrasts the picture that the surveyor John Hume drew of his Sutherland environment in 1774 to what can be seen today:

> [Hume] writes that grazings of Achmore in the parish of Assynt 'yield all the variety of sweet grasses to perfection'. Today the braes of Achmore rising to the north of Loch Assynt are a bare expanse of heather and deer grass with marks of erosion on the steeper faces. There is no sweet grass and little variety.[38]

Any keen eye can confirm the truth of such observations. Osgood Mackenzie contrasted Gairloch as described in his uncle's papers around 1820 with what it was a hundred years later:

The most perfect wild Highland glen . . . he says that no sheep had ever set foot in it; only cattle were allowed to bite a blade of grass there. The consequence was that the braes and wooded hillocks were a perfect jungle of primroses and bluebells and honeysuckle and all sorts of orchids including *Habenarias* [fragrant orchid] and the now quite extinct *Epipactis* [helleborine], which then quite whitened the ground.

Osgood Mackenzie himself had only seen the latter flower twice in the area.[39]

You have only to look at the mountain ledges, which the sheep cannot reach on Ben Lawers or Caenlochan, with their wealth of alpine herbs, or visit the exclosure on the National Nature Reserve at Inchnadamph with its slow but abundant recovery of dwarf willows, and contrast their profusion with the bare ground outside, to sense what has been changed by the nibbling sheep.

The present crisis in upland farming makes things worse, since the only profit to the farmer now comes from headage payments, increasing the temptation to get as many mouths on the hill as possible. Yet there are fewer than a thousand shepherds left in Scotland today. If the subsidy could be given for the shepherd rather than the sheep, and combined with payment for acreage rather than for the number of animals, a better and more traditional system could emerge, with a moderate number of good quality animals being moved from place to place to ensure that overgrazing never occurred.

The rise of the sporting estate is the second factor that has altered the character of the uplands in the last two centuries. The concept of the sporting estate was itself unknown before the nineteenth century, though the hunt was as old as human history. Nobles, abbots and kings had hunted on the moors with crowds of their retainers, closing in on the deer in a great circle – tinchel – or driving them into an enclosed area where they could be massacred in an orgy of blood with dog, spear and arrow. In parts of Sutherland they even drove them into the sea and killed them from boats. It was all very delightful from the perspective of the hunter. The Dean of Lismore in 1549, in the first systematic description of the Hebrides, cleric though he was, everywhere praised the chase. Thus in Islay, 'many woods, faire gaimes of hunting beside everey toune'; in Mull, 'maney deire and verey fair hunting games'; in Skye, 'maney woods, maney forrests, maney deire, fair hunting games, maney grate hills'; in Harris 'aboundance of deir, . . . very faire hunting games without any woods, with infinite slaughter of otters and matricks [pine martens]'.[40]

The nearest equivalent at that time to a sporting estate was a hunting reserve or 'foresta', a forest, though the term at least in Scotland never implied past or present tree cover. In royal forests and in those forests which were nominally royal but actually under the control of nobles in the early modern centuries, like Glen Finglas under the Earls of Moray or Mamlorne under the Earls of Breadalbane, husbandry might be modified to accommodate the pleasures of the chase. Moray intermittently fretted about how compatible his tenants' stock was with maintaining woodland for his deer, and Breadalbane had endless problems with cattle at the summer shielings. But landowners were fairly relaxed about their sporting rights before 1800. English tourists at the end of the eighteenth century who happened to bring their guns or fishing rods were usually welcome to take what game they chose provided they asked permission. The publication of the *Sporting Tour* of the Yorkshire gentleman Colonel Thornton in 1804 created a sensation in the south, with its tales of prodigious bags and catches in Strathspey and elsewhere. It went a long way towards opening the eyes of a very different class of Englishman from the Romantic poet and the aesthetic tourist to the opportunities for delight in Scotland.

Nevertheless, the idea of an estate the prime purpose of which was to entertain visiting sporting gentlemen had to wait for three Victorian developments. The first was improved accessibility to the hills by horse-drawn carriage and steamboat, and ultimately in 1863 by a railway that could bring the sportsman to Inverness after an overnight journey from London.

The second was a series of innovations for killing game – the 18-foot salmon-fishing rod of split cane and green-heart wood imported from the colonies,[41] the express sporting rifle, the breach-loading shotgun and, earlier and above all, the cartridge. The last named was invented in 1807 by Alexander Forsyth, an Aberdeenshire clergyman frustrated by the fact that sitting ducks were wont to take off because they could see the flash of his muzzle-loader before the shot hit them. Mr Forsyth, it must be added, was not so unworldly that he failed to see the military implications of his invention, which he developed in concert with the Royal Arsenal at the Tower of London. It has been described as the most important innovation in firearms since the original invention of gunpowder.[42]

The third development was the emergence of a vigorous land market in upland estates that enabled anyone who so fancied, in the rich elites of northern and southern Britain, to buy or rent a moor or a river for their own delight. This emerged in Highland Scotland after the spate of bankruptcies of older families in the mid-century crisis that ended in the potato famine, and more generally in the last third of the century when sheep farming began to suffer from foreign competition and to become a less profitable primary use of the land.[43]

By 1884, almost two million acres of the Scottish Highlands had

From the Langholm game book, 1893–4: 'opposite views.'
Duke of Buccleuch.

become deer forest. By 1912 it was about 3.6 million acres. At the latter
date, two-fifths or more of the land surface of Ross and Cromarty and
Inverness were devoted to the sport.[44] Further south and east – in Perthshire
and Aberdeenshire – the emphasis was on grouse moor rather than deer
forest; these moors still carried substantial flocks of sheep. In northern
England – Lancashire, Yorkshire and the Pennines generally – it was
exclusively on grouse, though some sheep might be kept. It is commonplace to attribute the passion for the sporting estate and its baronial
shooting lodges to the example of Queen Victoria and Prince Albert at
Balmoral. Their enthusiasm certainly helped, but the move had begun to
gather pace before Balmoral was built and was as obvious on the Yorkshire
moors or the counties on both sides of the Scottish border as in the
Aberdeenshire glens.

The grouse moor and the deer forest between them changed a landscape
of use, full of farmers working the hills at field and shieling, to a landscape
of delight, kept empty of people. But it was delight for a very small number
of rich owners and their friends, zealously guarded, as we shall see in due
course, from the incursions of the common herd.

One can easily form the impression that the worship of the hunt made
some of its votaries slightly unhinged. In 1872, for example, on Wemmergill
moor in Yorkshire 17,074 red grouse were shot in the season, of which its
owner, Sir Frederick Millbank claimed 5,568. A granite memorial was
erected not to the birds but to the achievement.[45] Lesser game got shorter

shrift and a briefer obituary. Lord Harewood noted in his shooting book for 1898: 'Terrible year for rabbits: 12,713'.[46] For grouse kills by one person in a single day, Maharajah Duleep Singh held the joint record of 440 birds shot over dogs on a Perthshire moor, but when it became fashionable about 1860 to drive the birds towards shooters concealed in butts, this was readily eclipsed. The world record for a single gun was set by Lord Walsingham in Yorkshire in 1888, when he killed 1,070 grouse with 1,510 cartridges at an average of 2.3 birds a minute, using four shot guns and assisted by two loaders and forty beaters. It was an organisational as well as a ballistic feat: no wonder one expert held that due to their knowledge of military drill, members of the volunteer rifle corps made the best beaters.[47]

The maintenance of moors to support huge populations of game, and also to have them available on demand for the few weeks in the year when the Victorian and Edwardian gentlemen took time to shoot them, demanded skilled and ruthless management. The skill on the grouse moor was in the correct burning of the heather in small patches, to ensure a mosaic of long old vegetation where they could hide and nest and short young vegetation where they could feed. The ruthlessness came in predator control. Lieut-Col Lord George Scott put it succinctly enough as late as 1937:

> To obtain first-class results with grouse on a moor, two things are absolutely necessary: a large area of really good heather and a total absence of vermin . . . Vermin must be regularly destroyed. Crows are bad vermin. Where rooks are too numerous they become a menace . . . Of course everybody knows about stoats, weasles, ravens, eagles, hawks, foxes, etc., all of which help to keep down grouse and blackgame. Rabbits are always bad for game.[48]

Here the historian is faced with an evidential problem. There is no doubt at all that enormous numbers of predators were killed by gamekeepers in the nineteenth century, for which figures are often quoted. Most come from James Ritchie's *Influence of Man on Animal Life in Scotland*. Ritchie was an excellent scholar but does not provide footnotes, so it is not possible to trace his sources. One comes from a newspaper of the mid-nineteenth century reporting on the contents of a vermin book, but again the original cannot be traced. But here are some of the figures from Ritchie: in five Aberdeenshire parishes clustering around Braemar, 70 eagles and 2,520 hawks and kites were killed between 1776 and 1786; on the Sutherland estates of Langwell and Sandside, 295 adult eagles were killed between 1819 and 1826; at the Dumfries fur market 400 polecat pelts were sold in 1829 and 600 in 1831, so that they became 'a drug on the market' – but the figure fell to twelve in 1866, and from 1869 there were no more polecats left to be offered at the fair. The newspaper report relates to Glen Garry in Inverness-shire, and claims in three years between 1837 and 1840 the deaths of well over 1,000 kestrels and buzzards, 275 kites, 98 peregrine falcons, 78 merlins, 92 hen harriers, 63 goshawks, 106 owls, 18 ospreys, 42 eagles and sundry other hawks in addition to some 650 pine martens, wild cats, polecats, badgers and otters. Another report speaks of 310 hen harriers being killed on Lord Ailsa's estates in Ayrshire between June 1850 and November 1854.[49] These figures are so extraordinary that I would dearly love to lay my hands on the originals.

Such gamebooks detailing vermin that I have been able to check in manuscript myself in the Scottish Record Office and elsewhere have seldom shown such dramatic kills, though some are striking enough. On the Breadalbane estate a single keeper over the fourteen years 1782–96 accounted for 131 polecats, 50 pine martens, 37 badgers, 45 weasels and 15 wild cats. The proportions are interesting.[50] More dramatic, and perhaps showing the benefit of cartridges, were the kills at Drumlanrig and Bowhill in three years from April 1829 of at least 688 hawks and 132 polecats (see Table 5.2).

Table 5.2 Vermin killed at Bowhill (B) and Drumlanrig (D) 1829–31

	April 1829–30		1930–1		1931–2		Totals
	B	D	B	D	B	D	
Hawks	381	41	96	16	[?]55*	15	688 +
Crows	199	119	240	98	122	85	863
Magpies	5	51	20	59	12	61	208
Weasels**	791	130	149	37	99	16	1,222
Polecats	49	19	19	11	15	19	132
Cats	69	34	46	33	29	44	255
Foxes	–	3	–	8	–	–	11
Hedgehogs –	–	–	–	103	–	103	

* The first digit is torn away
** Weasels – the category probably includes stoats
Source: Buccleuch muniments, Scottish Record Office: GD 224/519/366/2

Later in the century, when the original abundance of predators had evidently been vastly reduced, it was still possible in Glenshieldaig Forest between 1874 and 1902 to account (among other things) for 208 otters, 12 polecats, 33 buzzards, 63 kestrels, 17 golden eagles and a sea-eagle, 6 peregrines and 14 merlins. And on the Gruinard estate between 1904 and

1912, the water bailiffs shot 63 'water ousels' or dippers, a bird now known to be quite harmless to any sort of fish.[51]

Part of the 'general return of vermin killed' for seven years at the end of the nineteenth century, covering the very large and varied spread of Buccleuch estate properties at Dalkeith, Bowhill, Commonside, Newlands, Langholm, Drumlanrig and Sanquhar, is shown in Table 5.3.

Table 5.3 Vermin killed on the Buccleuch estates, 1894–1900

	Foxes	Weasels*	Cats	Hawks	Crows	Magpies
Total	427	6,383	4,327	1,341	7,596	764
Average per year	61	991	618	192	1,085	109

* Probably includes stoats

Source: Buccleuch muniments, Drumlanrig Castle.

Compared to early in the century, weasles and stoats had declined in relation to feral cats, and polecats have disappeared. Corvids (crows and magpies) were now very much more numerous as surviving predators than hawks. Surely one of the unintended results was to produce a world safe for rabbits.

Though one suspects that some of the totals in the vermin returns may occasionally have been inflated by the methods of paying bounties, the devastation that the gamekeeper armed with shotgun and gintrap did is unquestionable. By the outbreak of the First World War, the white-tailed eagle, the osprey, the red kite and the goshawk (all once common) were extinct in Scotland and northern England, the hen harrier survived only on some islands, the polecat was extinct and naturalists feared for the survival of the wild cat and the pine marten: virtually all other predators (except crows and perhaps the fox) were much rarer than they had been. Several of these species have staged a comeback in the twentieth century in a more favourable public atmosphere towards nature conservation, some, like the osprey, hen harrier and pine marten on their own accord, others like the white-tailed eagle, goshawk and red kite reintroduced with a little help from their friends. But the implacable hostility that owners of grouse moors continue to show towards hen harriers and peregrines, and the high level of illegal persecution of these and other protected species of raptor, continues to limit recovery to the full extent of their former range. Whether, given the extent of ecosystem change on the moors, raptors could ever remotely recover their former numbers within that range is extremely doubtful. Their previous abundance is itself evidence that the birds and small mammals which were their prey must once have been plentiful on a scale now barely imaginable.

In respect to predators, then, the sporting estate has, historically, seriously impoverished the uplands. On the other hand, the fact that so much land was devoted primarily to sport rather than to sheep could be expected to have had a cushioning effect on the uplands. Sheep were not banished from grouse moors, but good managers checked their numbers because of their tendency to reduce heather. Lord George Scott considered that if a proprietor 'can keep the sheep stock in his own hands, he can manage his grouse moor with ease, provided he takes personal interest in the heather burning and draining.'[52] Nor were sheep banned from most deer forests, but it was expected that most of the herbiage should be available for deer, not domestic stock, and on some forests – for instance, Rothiemurchus – they were not tolerated at all.

Certainly a well-run grouse moor, burnt in the traditional pattern of small mosaics, remains a much better home to most upland breeding birds than an intensive sheep farm although, since the Second World War, many grouse moor owners have allowed far more sheep on their land than would have been tolerated before.

The traditional deer forest also harboured a great deal of wildlife: golden eagles, for example, were favoured because they killed the grouse, whose cries startled the deer as the stalker approached. But not all deer grounds were kept pristine even in Victorian times. Experts on their management in the late nineteenth century criticised owners who destroyed heathery places and rough woods to grow grass for the deer, only to find the animals could not get at it when the snow came, or who drained marshes that grew cotton sedge in early spring: 'for some weeks it is probably the only fresh food the deer can get'. They also warned against feeding deer with hay or turnips: 'a great mistake unless *absolutely necessary*'.[53]

The deer forest of the late twentieth century is, unfortunately, often a less benign environment, as overstocked with animals as a sheep farm. In the eighteenth century, with a rising population of people and domestic animals, red deer were under pressure, seldom seen by tourists and frequently chased off by farmers. Under the protection of the sporting estate, their numbers began to grow in the nineteenth century. According to one estimate, they then doubled between about 1900 and 1940, to around 250,000. Under the pressures of wartime neglect and postwar poaching they fell by perhaps 40 per cent, only to recover to stand in 1991 at an all-time high of about 270,000 living on the open moor, plus an unknown number in forestry plantations. Recent estimates have been as high as 350,000.[54] Failure to cull the hinds, in the mistaken belief that more hinds would

The Caithness Flows: probably the largest blanket bog in the world.
Copyright S, Moore, Scottish Natural Heritage.

result in more trophy stags, and the growing practice of winter deer feeding (despite that wise Victorian warning), meant that animal densities rose
rapidly and had much the same effect on vegetation as overgrazing by
sheep. A density of four or five deer per square kilometre is held to be the
maximum that would permit the natural regeneration of trees in north-east
Scotland. When in 1996 the National Trust for Scotland took over the
Mar Lodge estate with a view to recreating the Caledonian pine woodlands
of the ancient forest of Mar, the density was at least three times that. In
1955 Fraser Darling reckoned that 60,000 red deer would have been the
ecological optimum.[55] That we now have five or six times that optimum
measures the influence of the Scottish landowner compared to that of the
scientific ecologist.

Our third development affecting the uplands, aerial pollution, must be
dealt with more briefly, which is not to imply that it has less significance.
Because of heavier rainfall and patterns of precipitation, aerial pollution is
often more serious on high ground. The fragility of the uplands is also
perhaps greater than that of many habitats at lower levels.

A historian accustomed to working with documentary sources may well
marvel at the variety, ingenuity and evident precision of scientific method
used to reconstruct the pollution history of the uplands in the last four
hundred years, ranging from the study of chemicals recovered from plant
material in museum collections, to that of carbon isotopes embedded in

old tree-rings, to that of the larval head capsules of different species of non-biting midges found in lake sediments in a series running back for three millennia, to similar work on fossilised diatom remains, and on particles of fly ash and trace metals also found in deposits of mud.[56]

The evidence from Lochnagar in the high Cairngorms is particularly impressive, since it would be difficult to find anywhere on the British mainland further removed from the sight and sound of urban industrial society – if there is utter wilderness in Britain it must exist in these high corrie lochans, unique in the United Kingdom for their arctic-alpine character. An increase in lead particles in their sediments begins to become evident from a level that probably represents a date a little before 1650, perhaps traceable to smelting round Leadhills, one hundred miles away, but more probably from the Pennines, 150 miles and more to the south. Lead pollution increased fourfold to a peak around 1900 and has dropped somewhat since. Zinc particles increased from around 1800 on a similar scale. Carbon particles, a main indicator of coal smoke and other industrial emissions, begin to accumulate around 1850, accelerated very rapidly in the twentieth century until around 1970, and have dropped by 40 per cent since, perhaps indicating the efficacy of the clean air legislation as smoke control orders spread in the decades following the legislation of 1952. Acidification itself can be measured from the changing composition of the diatom flora of the lochans, which shows instability from around the 1850s

The Caithness Flows: plough lines for forestry.
Copyright S. Moore, Scottish Natural Heritage.

but little alteration until around 1890, after which there was rapid change in favour of the acid-tolerating species which now dominate the waters. The implications of this study, of course, go far beyond the effects on the lochans themselves. It is clear, for instance, that the long-lying snow-beds on the Cairngorms are now much more acidic and more nitrogen-rich than formerly, though we cannot tell what impact this has had on the life of the snow-bed communities.[57]

Another interesting investigation has been of the variation of nitrogen content in *Racomitrium* moss, an important component of montane heaths in north-west Britain and one which has significantly declined in extent and quality in recent decades. Here the scientists found that moss from mountain summits near to urban centres in northern England has today up to sixfold more nitrogen deposited than moss in the far north-western Highlands, but that moss in nineteenth-century *herbaria* not only had appreciably less nitrogen, but the regional difference was then only about twofold.[58] As a generalisation it is fair to say that everywhere is polluted, but some places are much more polluted than others.

Much scientific effort has also been expended to discover what the effects of pollution have been in the past and to predict what might be in the future. We noted before that there had been big changes in the flora of the Pennines since 1800. What contemporaries called 'the reek of smoak of a lead mill' already led in the eighteenth century to frequent complaints of loss of stock and pasture, and when the great cities of Lancashire began to pour out sulphur dioxide in their coal smoke, naturalists (by the 1860s) noted a correlation between the local disappearance of certain lichens and bryophytes with atmospheric pollution. By 1913, the present extremely poor plant diversity of the southern Pennines was being compared to the richness of the eighteenth-century record and questions asked about the causes. The disappearance of *sphagnum* (once so dominant) was discovered in 1964 to be correlated with the arrival of soot in the peat profile, and although sulphur dioxide concentrations have declined in the past few decades, nitrate deposition has increased fourfold in the last 120 years, and this has been shown experimentally to prevent any return to the original moss vegetation. The southern Pennines, in fact, have demonstrably been seriously damaged by pollution for at least 150 years.[59]

Recent research has been concerned with the relationships between aerial pollution and the decline of heather. The UK average deposition of nitrogen has grown fourfold this century, with much of the Scottish uplands now receiving over 25 kg per hectare per year and much of Cumbria and the Pennines in excess of 30 kg.[60] Dutch researchers consider that there is a threat to heather moorlands at any level over half that, and while it has been argued that this is not necessarily true in Scottish conditions, experiments have shown that if the moor is grazed by sheep at the same time as it receives these heavy inputs of nitrogen, *Nardus* grass will indeed succeed over heather. A late twentieth-century combination of aerial pollution and excessive density of animals therefore appears to be peculiarly fatal to heather moor.[61]

Finally, those birds. I suggested above catastrophic falls since the nineteenth century in upland birds such as wheatear, dotterel and ring ousel in addition to the widespread decline in red grouse. Habitat change could of course be to blame, with the decline of heather moor and of montane heath and scrub. But research published recently suggests a possible additional factor. Over the British Isles since the 1850s there has been an appreciable thinning in the eggshells of blackbirds and thrushes, even when the effect of organochlorines in the 1960s is allowed for, and birds with thin eggshells may not be able to breed. If this finding is associated with research in the Netherlands showing that small birds have difficulty in bringing off clutches where calcium is in short supply, and evidence that in Scotland calcium on moorland soils declined between around 1960 and 1988 due to continuing acidification, it is possible to suggest a hypothesis that birds on the moors have been at a peculiar degree of risk for a century and a half from aerial pollution.[62] But this, as academics are so fond of saying, will demand further research.

CHAPTER 6

THE QUARREL OVER
THE COUNTRYSIDE

In the late twentieth century, the meaning, function and control of the countryside is contested ground. London has been the city of demonstrations and Hyde Park their focal point from the Chartists of the 1830s to the CND of the 1950s. Few have been larger, and none more conservative, than the gatherings there of the Countryside Alliance in 1997 and 1998. Of the first, in July 1997, *The Daily Telegraph* observed that had some disaster overwhelmed the park that summer afternoon the gene pool of upper-class, fair-haired, blue-eyed English men and women would have perished in its entirety. The object of that protest was to complain that the town no longer understood the country, the immediate occasion being to resist proposals in a private member's bill to ban the hunting of wild animals with hounds. Behind it, as became even clearer in the second demonstration six months later, lay inchoate anxiety aroused by fear of unparalleled rural depression, a threat which the previous government had seemed to exacerbate by its handling of the BSE crisis, and the new government, swept to power by an overwhelming urban vote, seemed likely to ignore.

It takes an effort to recall how, in the aftermath of their landslide victory, Tony Blair's government still glowed pink in tooth and claw. It was considering finding government time for the detested anti-hunting bill. It had vowed to abolish the political privileges of hereditary peers in the Lords, and in Scotland to undertake land reform of an unspecified but presumably radical kind. It was threatening to introduce legislation granting to urban ramblers the long-sought 'right to roam' in the countryside. It was considering a new Wildlife and Countryside Bill to improve protection for nature conservation sites. It had promised to stop the cull of badgers suspected of carrying bovine tuberculosis. Barbara Young, then chief executive of the assertive and distrusted RSPB, had become a working life peer on the government benches. New Labour, new danger indeed.

The size of the two demonstrations took everyone by surprise, though not everyone was as impressed as the government, which, after the first, promptly backed away from supporting the hunting bill and renewed its support for the badger cull – early indications that it would be pale blue rather than pale pink but in all events pale and cautious. In Scotland, the

Herald newspaper pointed out that one in five of the population lived now in the countryside broadly defined, but that only 2 per cent of the population were employed in agriculture, and of these only a miniscule number cared about fox-hunting or shared many of the other values of the Countryside Alliance. The correspondent suggested that as the rural population was growing again due to such factors as early retirement and homeworking (nowhere more so than in the Highlands), landowners and farmers were in part reacting against those from another culture who were coming to live among them.

More obviously, town people were coming into the countryside in unprecedented numbers as visitors, and signing up to join organisations that demanded a share in how the countryside was managed. The RSPB calculates that 18,000 people go bird-watching every weekend: it has a million paid-up members. The wildlife trusts have a quarter of a million members, the Ramblers' Association 130,000, the Woodland Trust 60,000. There are 22 million day-visits a year to the Peak District, more than to any other national park in the world except Mount Fuji.

The economic significance of these activities is immense. The value of tourism to the Scottish economy is estimated at £2.5 billion annually, greater than that of agriculture or forestry; it employs 180,000 people, involves 13.2 million trips and 70 million bed-nights a year. Visitor surveys show that the prime attraction of Scotland is 'scenery'; the key words are 'wide open spaces' and 'rugged landscapes', along with 'freedom, emptiness, isolation, peace, loneliness, variety and unspoilt countryside'. More specifically, in 1991 it was estimated that informal countryside recreation in Scotland generated an income of over £300 million and supported the equivalent of 29,000 full-time jobs. In 1996, Scots supposedly spent about £1,000 million on countryside day-trips, and UK residents spent over £250 million on activity holidays in Scotland. Hill walkers and mountaineers alone are credited with adding £150 million to the Highland economy in 1995.[1]

Though some of these figures are derived from questionable methodology, delight has become use in a big way. In all its forms from mass bird-watching to going on coach-trips from one hotel to another, countryside visiting has the potential to damage the natural world almost as much as more traditional occupations. Max Nicholson, former director of Nature Conservancy, nevertheless boldly argued in 1970 that the true significance of tourism lay 'in its direct and intense relationship to the appreciation of the natural environment, and in the emergence of large numbers of people interested in the environment for its own sake, and not merely as something to exploit.' All it needed, he maintained, was better education to

Demanding the right to roam: G. H. B. Ward, ramblers' leader,
at the Access rally at Winnats Pass, 1931.
Ramblers' Association.

understand what they came to admire.[2] The Countryside Alliance generally puts it in another way, and more tersely: the townie sees the countryside as a playground, not as a workplace. As William Cowper elegantly expressed it in the eighteenth century:

> He likes the country, but in truth
> Most likes it when he studies it in town.

Planning the National Parks: John Dower, c. 1944.
Michael Dower.

The demonstrators in Hyde Park, with tweed cap and Labrador, were complaining that urban critics were ignorant busybodies, interfering in the way farmers and landowners went about their business, whether it was farming, forestry or field sports. Yet neither side contested the need for interference. On the one hand, there is an extremely serious crisis in agriculture which to the farmer demands solutions that will at least secure his income and save him from being driven off the land: 'Keep Britain farming' in the words of the National Farmers' Union's current slogan. On the other, questions are asked about the return that the public is getting from a social group whose income and way of life has rested on subsidy and price support for five decades. It is an argument about what sort of interference is appropriate and what its limits should be. It is also, therefore, a quite fundamental argument about property. There is a gradually increasing expectation of the urban population that the countryside will be managed in a way that takes account of their interests. There is increasing resentment by landowners and farmers that such expectations will restrict their ability to manage their own property as they wish, and will impoverish them.

This argument has roots that go back to the Industrial Revolution. In 1810 William Wordsworth described his beloved Lake District, in his much read and reprinted *Guide*, as 'a sort of national property in which every man has a right and interest who has an eye to perceive and a heart to enjoy'.[3] A sort of national property! The words at first sight ring hot and dangerous twenty years from the French Revolution and the Reign of Terror. Yet Wordsworth, for all the apparent radicalism of that statement, was really talking about the interests of an elite of connoisseurs, 'persons of taste' trained to develop that 'eye to perceive' and 'heart to enjoy' by following the Romantic code of aesthetic observation of which he was high priest. As his attitude in the 1844 campaign against the railways with their expected trippers demonstrated, he certainly did not want or envisage the common crowd of labourers, clerks and tradesmen to claim such a right. For them, as Dorothy said, 'a green field with buttercups' should suffice.

But already in the Wordsworths' lifetime there was a vigorous plebeian tradition of appreciation of the countryside, from which questions gradually also began to be asked about property rights. We have already noted the intense delight in the mountains and their deer in the poetry of the Highland gamekeeper, Duncan ban Macintyre, paralleled in the Lowlands by Robert Burns's sharp farming eye for a mouse's nest, the fall of the snow on the river and the drop of poppy petals. Raymond Williams has shown how in southern England the poems of George Crabbe (son of a customs officer) and John Clare (farm labourer, gardener, lime-burner) were a counter-weight to the studied Arcadian tradition where the harvests were always golden and the squire always beneficent – and Clare, in particular, had an achingly realistic eye for the beauty as well as the hardship of the country-side.[4] These were not unusual men except in their articulate ability to turn their experience into literary form. The enjoyment of the natural world through fishing, catching animals and birds, egg-collecting, botanising, various sports, walking and simple nature observation was widespread among the same groups in society who agitated for Parliamentary reform.

None were more radical in the 1790s than the weavers of Paisley, who had their own botanical society and raised and read nature poets like

Robert Tannahill in their own ranks. From their community came Alexander Wilson, who fled political persecution to become America's greatest ornithologist in the generation before Audubon, painting and describing the great majority of the birds of the east coast.[5] A fine copy of his great nine-volume book resides appropriately in Paisley Museum.

Take, as another example, Thomas Bewick, the working Newcastle engraver whose art was unsurpassed in its direct observation of English animals, birds and rural sport. His autobiographical *Memoir* is redolent of a culture that delighted in nature. He recalled as a boy watching birds at the byre door of his father's farm, knowing his 'intimate acquaintances', the robins, wrens, blackbirds, sparrows and crows, but being moved to 'extreme pleasure and curiosity' when snow brought in woodcock, snipe, redwings and fieldfares. His lifelong passion, however, was angling, and, like most fishermen, he loved it for the surroundings as much as for the fish. He described the streams that:

> Run murmuring from pool to pool through their pebbly beds – and all these bounded and bordered with a background of ivey-covered ancient hollow oaks, with elms, willow and birch – thus clothed as if to hide their age, or offer shelter to an undergrowth of hazel, whins, broom, juniper and heather with the wild rose and woodbine and brambles, beset with clumps of fern and foxglove, while the edges of the mossy braes are covered with a profusion of wild flowers which 'blow to blush unseen' or which peep out among the creeping groundlings, the bleaberry, wild strawberry, harebell, violet and such like . . . how often have I, in my angling excursions, loitered upon such sunny braes, lost in extacy.

Possibly with a dig at Wordsworth, he went on: 'The man smit with a fit turn of mind and inclined to search out for such *beauty spots* will not nead the aids of poets to help him in his enthusiastic ardour.'[6]

Property came into question for Bewick because, like so many others, he was enraged by the highly successful efforts of the landed classes to extend what he called 'severe and even cruel' Game Laws to prevent the poor from taking a rabbit, a bird or a fish as they had traditionally done. In respect of angling he maintained that 'no reasonable plea can ever be set up, to shew that the fish of rivers ought to be the private property of anyone'. As for game, 'to convince the intelligent poor man, that the fowls of the air were created only for the rich, is impossible and will for ever remain so'.[7] The Highlanders' traditional conviction that three things in life were free: the deer on the hill, the fish in the stream and the tree in the wood, exactly parallel this mind-set.

Bewick was miles from being a revolutionary. He believed in the virtue of social consensus and strongly disagreed with his friend and fellow townsman, the bookseller Thomas Spence, who advocated the nationalisation of the land. Bewick held that 'property ought to be held sacred'. But a combination of considering the game laws and the Highland clearances, led him to a caveat:

> Property in every country ought to be held sacred, but it ought also to have its bounds and (in my opinion) to be in a certain degree held in trust jointly for the benefit of its owners and the good of society of which they form a part – beyond this is despotism, the offspring of misplaced aristocratic pride.[8]

This view was not only typical of his class in his age, but also represents what most people think about landed property today. The difficulty, of course, is in defining the respective limits of 'the benefit of its owners' and 'the good of society'.

The first to assert themselves against the view that rights of property in the countryside were absolute were walkers around the industrial cities, whose story Dr Harvey Taylor has told with admirable clarity.[9] The habit of urban people walking for pleasure and healthy exercise in the country was well established by the late eighteenth century. The remarkable journals of Adam Bald, dry-salter of Glasgow, tell of his rambles with his friends between 1790 and 1830, sometimes on walking holidays of five to fourteen days, often from a ferry stop along the Clyde coast, sometimes inland around Stirling, or from Dumbarton to Inveraray on the open hill. His was already the smug and hearty language of the fit. In 1791 he wrote that the seaside, once the monopoly of invalids, had become the resort of 'the plump and jolly, sauntering the rocky shore or climbing the heathery hill full of health and spirits, whilst the sickly race are confined to their gloomy chambers'.[10]

Significantly, it was around Glasgow, York and Manchester that the earliest clashes between walkers and landowners occurred. The background was the notorious 'Stopping-up Act' of 1815, which extended earlier legislation of 1773 to permit the closure of established rights of way if two magistrates signed an order declaring them 'unnecessary'. Fysche Palmer, one of the Act's critics, claimed that 'it was a common thing to hear one magistrate saying to another "come dine with me, and I shall expect you an hour earlier as I want to stop up a footway".' Such naked class legislation provoked a response among urban radicals. Pre-eminent among them was the Manchester newspaper proprietor Archibald Prentice, who claimed that 'thousands and tens of thousands whose avocations render fresh air and exercise an absolute necessity of life, avail themselves of the right of footway through the meadows and cornfields and parks of the immediate neighbourhood.'[11]

Disorder occurred first in 1822, at Dalmarnock bridge near Glasgow, where a landowner barricaded a path by building a spiked wall across it; the opposition was orchestrated by a local bookseller. In the ensuing riots which attempted its demolition, the military were called in and took forty-three prisoners. Only four were charged (a collier and three weavers), which set in motion a lengthy legal procedure culminating in 1829 in a House of Lords judgement in favour of the protesters and a declaration of a legal a right of access along the entire path.[12]

Similar events near York led to the formation of an Association for the Protection of Public Footpaths around that city in 1824, and, most important of all, uproar concerning footpath closure in Lancashire between 1824 and 1830 led to the formation of the aggressive and successful Manchester Footpath Preservation Society, skilfully orchestrated by Prentice, who personally led the way in levelling obstructive banks, fences and iron hurdles. In Dr Taylor's words, its membership was 'drawn overwhelmingly from those who had initiated and developed commercial and professional-based liberal radical dominance in the city', and it was led 'by the young, successful and largely nonconformist'. Several of the activists had earlier been involved in protest against the behaviour of the authorities at Peterloo, or lent support to the Blanketeers, and were becoming involved at the time in the agitations of the Anti-Corn Law League.[13]

The ensuing fuss led, significantly in 1833 (after the election of a Whig government and the passing of the First Reform Bill), to a Select Committee on Public Walks and two years later to a General Highways Act that modified the law to make it more difficult, but by no means impossible, for landowners to obstruct footpaths.[14]

Of course the conflict did not end here, and the campaign for public access in the countryside continued to fight its battles throughout the nineteenth century and beyond to the present day. In 1844 the forerunner of the Scottish Rights of Way Society was formed in Edinburgh by similar sorts of urban people to those who had supported the cause in Manchester, but with a distinctive input by lawyers. The civic leaders at the inaugural meeting complained that the well-being of the city was threatened by the surrounding landowners obstructing the footpaths, and that to reopen them would be a cheaper way to restore public health than the expensive measures to reform the sewerage proposed by Edwin Chadwick. It became a national as opposed to a municipal concern in 1847 when the Duke of Atholl's gamekeepers came to blows with an Edinburgh professor and his students on a botanical outing on the old drove road in Glen Tilt, and the

*The summit of Ben Nevis in 1885, over a century before
plans for a Cairngorm funicular.
St Andrews University Library, Valentine Collection.*

society (with the support of Perth Town Council) fought him all the way to the House of Lords.[15]

The following year John Stuart Mill published his *Principles of Political Economy*, making a case that ownership of land was conditional on respect for the public interest, and in particular denying that rights of property over open land included the right to deny access to others. He went on:

> The pretensions of two Dukes to shut up a part of the Highlands and exclude the rest of mankind from many square miles of mountain scenery to prevent disturbance to wild animals is an abuse. It exceeds the legitimate bounds of the right of landed property. When land is not intended to be cultivated, no good reason can in general be given for its being private property at all.[16]

It was an important endorsement of the popular view.

From this point onwards, the emphasis in northern Britain turned to defence of access to and through the uplands, whereas in the south it became more concerned about keeping open the commons such as Hampstead Heath and Epping Forest, as well as about encroachment on paths and highways. In Scotland, the Scottish Rights of Way Society (among many other cases in the last quarter of the century) successfully defended Jock's Road up Glen Doll against Duncan Macpherson, a new landowner who had made his pile in Australia. The legal costs ran to over

£5,000, which they largely recovered after the House of Lords found in their favour, but it was a big risk for a voluntary society. Similarly they defended the Lairig Ghru between Strathspey and Braemar against the old landed family of Grant of Rothiemurchus and other roads in Deeside against established families.[17]

The whole business of access flared up around 1880 with the growing importance of sporting estates, and an anxiety by shooting tenants, in particular, to avoid their fun being disrupted by deliberate or accidental disturbance by walkers. Most famous was the absurd lawsuit brought by an American millionaire and shooting tenant, W. L. Winans, against a crofter, Murdoch Macrae, for the trespass of a pet lamb on the grazing of his 200,000 acre deer forest, dismissed by the Court of Session in 1885 with costs against Winans, but the defence had been organised by radical liberal elements in the legal profession and funded by a Highlander who had emigrated.[18]

This was the background against which the Liberal MP and later cabinet minister, James Bryce, introduced his series of unsuccessful Access to Mountains (Scotland) bills on eight occasions between 1884 and 1909, declaring that no owner of uncultivated mountain or moorland should be entitled to exclude any person from walking on such lands for the purpose of recreation or scientific or artistic study. Originally designed only to

The A9 at Aviemore in 1939.
St Andrews University Library, Valentine Collection.

apply to Scotland, Charles Trevelyan in 1908 introduced a similar bill to extend such provisions to the rest of the United Kingdom.[19] Almost without change, they form the foundation and inspiration for the so-called 'right to roam' which the Ramblers' Association has urged upon unwilling governments down to the present day.

There are a number of significant points about nineteenth-century access to the countryside that deserve emphasis. Firstly, in northern Britain very large numbers of urban people, both working class and middle class, were involved, many of them in militant campaigns. In 1877, a demonstration with an estimated attendance of between 20,000 and 40,000 forcibly removed a railway line that a mining company had erected on Hunslet Moor near Leeds. The upshot was that the moor became enshrined in statute as common land, to be 'preserved as open spaces for public use or recreation'. Some outings of the Victorian naturalist clubs attracted numbers comparable to bird-watching twitches of the present day: the Manchester Field-Naturalists Society recorded an attendance of 550 on one such outing. A survey of 1873 uncovered 104 field clubs in Britain, and another made at the end of the century estimated the combined membership of all natural history societies at nearly 50,000.[20]

Hill sports, too, increasingly had their devotees, as it became appreciated that alpine climbing could be enjoyed closer to home. In the 1890s, the Scottish Mountaineering Club had a conciliatory culture and endeavoured to work with landowners to seek permission to climb (and was seldom refused). The Cairngorm Club was confrontational and radical, and trailed its coat by deliberately challenging stalkers, landowners and their ghillies and publicising the confrontation in its *Journal*.[21] The Clydesdale Harriers were typical of a local organisation, with 700 members and mass runs on the hill of fifty young men together. They usually sought permission, but sometimes inadvertently disrupted a stalk or a shoot.[22] The first volume of the *Scottish Cyclist* was published in 1888, of the *Scots Angler* in 1896, of the *Scottish Ski Club Magazine* in 1909. The club itself had been founded in 1907 and had a fund to give away skis to postmen and shepherds for use in their work. They claimed that miners in the north of England had used something like skis a hundred years before.[23]

Far more numerous than those who belonged to the official clubs, however, must have been those who enjoyed the countryside simply as a place to walk and cycle. As early as the 1890s the Church of Scotland was complaining that cyclists were using their machines to cycle past the church on Sundays, not to cycle to it: their jollification in the countryside upset the Sabbath.[24] The hill-walker and the rambler were also often a conspicuous element on excursion trains and boats, sometimes described with condescending amusement by the social superiors. Thus a traveller on one of the Loch Lomond boats around 1883 described a host of motley clerks and warehousmen

> so transmogrified by the assumption of knickerbocker or kilt suits that the mothers who bore them wouldn't have known them, as they appeared in their tourist rigs to enjoy their hard-earned holidays. There they were, in all the glory of tacketted boots, and Tam o' Shanter or Glengarry bonnets, with knapsack on back and stout cudgel in hand, going to 'do the Ben', or the Trossachs or perchance prosecute a walking tour from the head of the loch to stern Glencoe.[25]

They were as strange as Hindus to his upper-class eye.

Next, use of the open hill was a phenomenon quintessentially northern British. The heartland of the walking, rambling and climbing clubs was the industrial towns of Lancashire and Yorkshire, and of the central belt of Scotland. The moors were too far from London, Birmingham, Bristol or Norwich to be of any interest to working men there, though use of commons, rights of way, and river banks and canal banks for angling clubs was of course important.

Thirdly, and emphatically, the contest was often seen in terms of town versus country, the people versus the landlords, Liberal versus Tory, and considerations about the meaning of property were drawn into it which echo the common view articulated by Thomas Bewick and endorsed by Mill. Thus James Bryce in the debate on his access bill in 1892 almost paraphrased Mill:

> Property in land is of a very different character from every other kind of property. Land is not property for our unlimited and unqualified use. Land is necessary so that we may live upon it and from it, and that people may enjoy it in a variety of ways; and I deny therefore, that there exists or is recognised by our law or in natural justice, such a thing as unlimited power of exclusion.[26]

Bryce, though, was open to the criticism levelled by J. Parker Smith (a parliamentary opponent, yet himself a supporter of the Scottish Rights of Way Society) that the access bill totally ignored local country people who needed local paths most to get to work or to travel on their daily business. It was for tourists on holiday, artists and scientists – that is, urban outsiders whom, it was assumed, contributed little or nothing to the local economy.[27] That it was a town-dweller's claim on the countryside made it doubly offensive to those who resisted it.

It is remarkable how little the objectives and the arguments on either side have changed since the end of the last century. The very titles of Marion Shoard's books, *This Land is Our Land* (1988) and *The Theft of the Countryside* (1980) reveal the ancient ideological base of the Ramblers' case. Proponents are still seeking to achieve James Bryce's aim of unfettered access to open and uncultivated land through legislation, as a matter of common justice and social welfare. Ten further bills were promoted by his successors before the Second World War, and, following the unsatisfactory Access to Mountains Act of 1939 and the National Parks and Access to the Countryside Act of 1949, which both failed to accede a general right of access, several further bills were introduced thereafter, as in 1978, 1980 and 1982. The only fresh argument heard is to invoke comparison with the law in Europe, notably the so-called *allemänsrett* of Norway and Sweden.

Opponents similarly still argue that existing access law gives freedom enough, and that to extend it into a general right would invite disruption of country activity by ignorant or loutish town-dwellers. The ramblers still tear down illegal obstructions on footpaths as they did in Victorian days, and the most famous twentieth-century demonstration, the so-called mass trespass on Kinder Scout in 1932 (which actually did not go near Kinder Scout, and involved a punch-up between eight keepers and up to 200 militant ramblers from the British Workers Federation a hundred yards from a footpath) had distinct nineteenth-century overtones.[28]

But things are changing. Initiatives designed to bring walkers and landowners to the same table, notably Scottish Natural Heritage's Access Forum of 1994, and the subsequent 'Concordat on Access to Scotland's Hills and Mountains' of 1996, signed by ten participating bodies, were initially more warmly welcomed by landowners than by ramblers: the former saw them as a way to stave off legislation, the latter as potentially trapping them into compromise on a matter of right. Now, however, both in Scotland and England, government proposals for legislation are coming close to acceding a right of access. In Scotland, they are less feared by landowners and farmers and less distrusted by walkers and mountain climbers, precisely because five years of sitting round the table together has helped to change attitudes. In England, the tone of Parliamentary debate is as confrontational as ever it was, though the government bill is likely to come close to acceding to the ramblers' demands.

Another and altogether less confrontational way of securing the land for the people was to buy some of it on their behalf. This was the strategy of the doughty inheritor of William Wordsworth's defence of the Lake District, Canon Hardwicke Rawnsley, and his allies Robert Hunter,

Octavia Hill and the Duke of Westminster. Their campaign to save beauty spots that were coming onto the market, such as the Lodore Falls south of Keswick, led in 1895 to the formation of the National Trust for Places of Historic Interest or Natural Beauty. Transformed by Act of Parliament within twelve years into the modern National Trust, it became a corporate body holding land inalienably on behalf of the nation, and was then imitated by the National Trust for Scotland in 1931.[29]

The concerns of the three founders other than Rawnsley went much wider than protecting the Lakes. They focused, like Octavia Hill, on the need of the urban poor of London and other great cities for access to open space and natural beauty. Yet the prime purpose of the National Trusts has never been to promote unrestricted access but to secure the preservation of selected and particular fine buildings and landscapes for the general good. It uncannily mirrors in its selectivity the different world-views of a Bewick and a Wordsworth. Nevertheless, the view of property held by its founders was recognisably similar to that articulated by James Bryce at the same time. Here is Canon Rawnsley in the *Contemporary Review* of 1886, complaining that, in the Ambleside neighbourhood alone in the previous fifteen years, twenty-one supposed ancient rights of way had been closed against tourist and neighbour alike. After quoting Ruskin's opinion that 'of all the mean and wicked things a landlord can do, shutting up his footpath is the nastiest', Rawnsley continued:

> That property in land has rights, responsibilities and restrictions of its own, none will at this time of day deny. Not till this is fully realized shall we cease to have occasion to wonder, with the Duke of Westminster, 'at the mania for shutting out the public from everywhere'. Until then we shall still be apt to look upon a piece of property in land much as one looks at the possession of a piece of china or a consignment of cotton.[30]

Again, there are remarkable echoes from a century later. An article on land reform in Scotland in the autumn of 1998 begins: 'Landowners claim the land as their own, saying that it's just like you or me owning a chair or a house.'[31]

Where the two traditions diverge is about who defines, in Rawnsley's phrase, 'rights, responsibilities and restrictions'. There were and are, many landowners prepared to agree that possession of land is not like possession of china or a chair – that it carries a degree of public responsibility summed up as stewardship, and implies a duty to share the benefits of good stewardship with a wider community. That, after all, was the traditional High Tory view, used to counter the claims of radical Liberals of 'the land for the people'. Such landowners, like the Duke of Westminster himself, have

played a very significant part in both National Trusts since their founda-
tions as chairmen or council members, but they have never been likely to
emerge as leaders of the ramblers.

Of course not, for the National Trusts disarmed criticism of the landed
classes just when the ramblers began to fuel it. The Trusts' tradition
emphasises, in all their enormous holdings of land and stately homes, the
public benefits that have accrued in the past from landowners defining
'rights, responsibilities and restrictions' in their own way, and from their
present definition by the committees that run the property, often led still
by the titled. Houses and estates now in Trust ownership are invariably
presented as memorials to good stewardship, past and present. They are
held up as a mirror in which the public can identify the landed classes. The
two Trusts have been vastly more successful in membership terms than the
ramblers because they have had bargains to offer: a day out for the family
at a reasonable price. But the urban public have in their coach loads not
objected to the view in the mirror and have also identified with the portrait
that benevolently smiles back: there is always a frisson if the owner briefly
appears. The two Trusts have poured much oil on the troubled waters that
divide town and country. Considering the outcry that there was against
the House of Lords under the reforming Liberal government at the start of
the twentieth century, just as the National Trust obtained its charter, a
speculative case could be made for arguing that they delayed the abolition
of the voting powers of the hereditary peers for nearly one hundred years.

If the access lobby has produced no new arguments since the nineteenth
century, and the National Trusts have acted as an emollient rather than an
irritant in town–country discord, what has nevertheless fed the quarrel so
strongly in the twentieth century?

The first relevant factor is the town-and-country planning movement,
which began to get underway in the interwar years and reached its apotheosis under the 1945 Labour government. Its character was ambiguous. It
began as a crusade in the tradition of John Ruskin and William Morris, to
defend the rural and the vernacular from the gross impacts of modern
urbanism, in the interests alike of country and town. Two years after the
foundation of the Council for the Preservation of Rural England in 1926,
the architect Clough Williams-Ellis published his indictment of rural
degradation, *England and the Octopus*. G. K. Chesterton was ecstatic in
review: 'No more valuable warning has been put into words in our time.'
He went on to observe that the modern age was in peril of ignoring the
classical notion 'that town and country are two completely different things:
that each should have its totally different dignity'.[32] Ultimately the town-
and-country planning movement did reassert this dichotomy, bringing all

*Using nature: Inverpolly National Nature Reserve
from the fish farm, Wester Ross, 1989.
Peter Wakely, English Nature.*

sorts of benefits in slowing the erosion of rural beauty and character, yet in doing so it perhaps legitimised future claims by the country to be 'special' in respect to resisting demands made upon it by urban people.

England and the Octopus was, its author was later to claim, 'an angry book, written by an angry young man' (he was actually forty-four when it was published). It was certainly eloquent:

> Decent, God-fearing, God-damning Englishmen live very contentedly in the pink asbestos bungalows; and if they chance to [live] . . . where they can be seen from miles around, they are the more content.

He was against ribbon development along the new motor roads that stretched out from the towns into the country, where 'disfiguring little buildings grow up and multiply like nettles along a drain, like lice upon a tapeworm'. He was against advertisements in the country, badly placed radio masts and pylons and 'the arbitrary harshness of the huge pipelines laid down the mountainsides' by hydro-electric concerns with no attempt at concealment or camouflage. He was for National Parks, new planned towns, planning restrictions on building, tax concessions to save great houses – a visionary indeed.[33]

The sequel to *England and the Octopus* was *Britain and the Beast*, edited by Williams-Ellis nine years later (in 1937), with twenty-six distinguished contributors ranging from J. M. Keynes, E. M. Forster and G. M. Trevelyan to A. G. Street and Patrick Abercrombie, and with the endorsement, among others, of Ramsay Macdonald, Lord Baden-Powell, Sir Stafford Cripps, Julian Huxley and J. B. Priestly – a remarkable coterie of the great and the good, and of the future powerful. They railed against and promoted similar things that Williams-Ellis had railed against and promoted before, though there was notably more anxiety about the speed and extent to which agricultural land was going under bricks and mortar, and about the implication of the car with its inevitable handmaiden, the 'black, shiny, unsympathetic roadway' that was replacing the traditional unpaved country lane. Professor Joad wanted roads 'canalised' so that cars could not escape from arterial routes between towns. Particularly interesting, however, was Joad's opening observation:

> The people's claim upon the English countryside is paramount . . . [but] the people are not as yet ready to take up their claim without destroying that to which the claim is laid . . . [so] the English countryside must be kept inviolate as a trust until such time as they *are* ready.[34]

For all its paternalism, it embraced the Bewick tradition and indicated that not all who were interested in planning wished to declare apartheid between town and country and leave the latter undisturbed.

It is difficult in this debate to uncover what ordinary country people thought of all this. In an attempt to find out, I explored the minutes of the Yorkshire Federation of Women's Institutes between 1926 and 1933. This was the biggest federation of its kind in Britain, and representative at the time of a wide cross-section of rural opinion. In 1928 they complained to the county councils of the 'unsightly tin dumps which disfigure our villages, and the tar containers which are left in horrible heaps'. They suggested a recycling scheme for tin, already practised, they claimed, in some towns where it 'proved a useful source of revenue'. In 1931 they complained about the routing of pylons across the moors so that 'the beauties of nature [are] utterly destroyed'. In 1930 and in 1933 they were worried about the need for footpaths on country roads for children and others walking to school. One delegate proposed fighting for by-laws prohibiting motorists from exceeding a certain speed when passing through villages, but Mrs Johnson of Ingleby Arncliffe spoke up for the motorist:

Too many women pushed their perambulators in the road: she knew how unpleasant it was to meet such people especially at a bad corner.[35]

It must, of course, have been infinitely more terrifying to meet Mrs Johnson. This is the moment of Mr Toad, when the road is claimed from other uses and becomes what a Lakeland observer called the 'nightmare of rushing cars and motor coaches . . . Something will have to be done about them now unless holiday traffic can be restricted by law.'[36]

By the time of the Second World War, there was consensus across party lines that national land-use planning needed a much more effective bite than that afforded by the tentative and piecemeal legislation of the 1930s, a view shared as much by rural land-use experts like Dudley Stamp as by regional planners like Patrick Abercrombie. The vision for the future in England was set out by the reports of the Barlow Commission on the distribution of the industrial population in 1940, by that of the Scott Committee on planning in rural areas in 1942, by that of John Dower on National Park areas in 1945, and by that of the Hobhouse Committee on National Parks and Access in 1947. The main upshot for the countryside was the Town and Country Planning Act of 1947 that introduced the concept of a comprehensive system of development plans, and the National Parks and Access to the Countryside Act of 1949 that provided National Parks in England and Wales only, and a Nature Conservancy service in England, Wales and Scotland.[37]

The rhetoric of the period was almost in a familiar tradition. The Scott Report spoke of 'the countryside as the heritage of the whole nation . . . it is a duty incumbent upon the nation to take proper care of what it holds in trust'. The Hobhouse Report spoke of 'the principle that the heritage of our beautiful countryside should be held in trust for the benefit of the people'. The word 'heritage' appeared where an earlier generation might have said 'property', perhaps to allay widespread suspicions that National Parks would imply compulsory purchase and land nationalisation. Finally, Lord Silkin, Minister of Town and Country Planning, spoke in the debate on the 1949 Act of: 'A people's charter . . . With it, the countryside is theirs to preserve, to cherish, to enjoy and to make their own.'[38]

Reality, of course, was not remotely like this. Firstly, postwar legislation came no closer than before to conferring a general right of access over open country such as Bryce and his successors had sought. Secondly, because of what was seen as the overriding postwar need to produce food and timber from domestic supplies, agriculture and forestry were exempted in the 1947 Act from development control. It had been assumed there could here be no conflict between use and delight – as the Scott Report explicitly put it, 'there is no antagonism between use and beauty', and the landscape was 'a striking example of the interdependence between the satisfaction of man's material wants and the creation of beauty'. Therefore, even if there had been no economic, social or strategic considerations, 'the only way of preserving the countryside in anything like its traditional aspect would still be to farm it'. In 1940 it was still the traditional, labour-intensive predominantly horse-powered and organic farming that was in question, and this the Scott Report, sensing change, wanted to see maintained for social reasons. Even at the time *The Economist* wrote off that ambition as 'vague romantic flubdub'.[39]

Thirdly, despite parallel committees established by the Scottish Office under Douglas Ramsay, in the end no National Parks were proposed in Scotland. Partly this was due to opposition from landowners, who smelt in the ambitious proposals to buy land for the parks the whiff of rampant socialism. Partly it was due to opposition from the hydro-electric interest, who could not but be suspicious that National Parks would hamper their plans for ambitious expansion in the Highlands. Partly it was because local authorities quarrelled over where the boundaries might be and worried how a National Park Authority might trample on their interests. Partly it was due to the Forestry Commission, who from their first reading of the Dower Report feared (in their chairman's words) that 'Mr Dower had entirely missed the point that there was much more of interest in walking through a wood than over a bare hillside', and that the result of his recommendations would be to 'sterilise large sections of the countryside'.[40] Partly it was due to the attitude of the National Trust for Scotland, questioning the need for 'the high costs of land acquisition and the expensive and unwieldy administrative machinery'.[41]

The Forestry Commission and the NTS both had a vested interest in promoting their own Forest Parks and mountain estates (such as Glencoe) as alternatives to National Parks, especially as the perceived pressure of the urban population on the countryside was nothing like as great as in England. But the NTS had another reason. The Douglas Ramsay Committee, in an attempt to counter the sterilisation argument, went overboard in its 1947 report in emphasising how National Parks would bring in the tourists and be friendly to the extension of forestry and agriculture 'where nature has left bleakness'.[42] The Committee had been urged by the main planner among them, Robert Grieve, to think:

> In terms of development rather than of preservation; it is inevitable this should be so in view of the circumstances of the Highlands and their past history. Our greatest anxiety will be to see that there is life and prosperity in our National Park areas: life depending on more than tourism; life based

Professor Peacock of the University of Dundee calculates the
risk of walking in Glen Doll in 1935: 'Deer-forests are Private Property and
the Public are requested Not to Trespass from the Path which is Clearly marked.
To do so, in addition to ruining a day's sport, is Extremely Dangerous
as long range high velocity Rifles are in constant use.'
Dundee University Archives.

on an integration of agriculture, tourism, indigenous industries, forestry and hydro-electricity.[43]

Grieve was later to become, understandably and more appropriately, the first chairman of the Highlands and Islands Development Board. The Douglas Ramsay Committee did not totally agree, but they allowed Grieve to publish an addendum to their report, on planning in National Parks as exemplified by what could become of Glen Affric, very much in favour of hydro-electric development in one of the most celebrated glens in Scotland: 'the dams and power houses should be of great interest to visitors'.[44]

It was not therefore unreasonable of the NTS to express doubts about the Committee's ethos, to note that some of the members 'visualise the creation of development areas rather than what most people would term National Parks', and to wonder if their enthusiasm for tourism would lead them to set up golf courses and tennis courts.[45] With so many enemies and no friends, the proposals were shelved for half a century. National Parks will no doubt come to Scotland early in the new millennium as part of Labour's proposals before the Scottish Parliament. Perhaps there are now rather different preconceptions as to their character and purpose, but the arguments that Grieve was raising have not been forgotten. In their advice to government, Scottish Natural Heritage suggests that National Parks in Scotland should have 'a purpose of social and economic development' as well as of protection and enhancement, but with 'the balance of interests favouring the long-term protection of the natural resources'. Experience elsewhere suggests that balance will not easily be struck.[46]

England, of course, did get her National Parks in 1949. All ten of those in existence by 1980 were moorland parks, demonstrating the hold of the old northern English tradition on the activists close to the levers of power. John Dower himself was born in Ilkley and lived in Northumberland: he warned, 'watch my bias to the North Country'. His wife's uncle, the noted historian G. M. Trevelyan, was one of the most powerful polemicists for countryside preservation of his generation. He believed that mountains did people more good than lowlands, having 'more rugged strength and faithfulness with which we can converse' than the effete woods and smooth hillsides of the south. Such romantic and lofty preferences were shared by Tom Stephenson, leader of the Ramblers, who managed to get the Hobhouse Committee to remove the Norfolk Broads and the South Downs from their list of potential National Parks on the grounds that, despite their popularity, they were not wild enough.[47]

The Dower Report declared the two 'dominant purposes' of English National Parks to be that 'the characteristic beauty of the landscape shall be preserved', and that the visiting public should have 'ample access and facilities within it for open-air recreation and enjoyment of beauty'. There was no talk then of a purpose of social and economic development, but it was not intended either that any obstacle should be allowed to agricultural improvement. In the atmosphere of wartime and postwar food shortages, such was unthinkable. 'Efficient farming', said the Report, 'is a key requirement in National Park areas', and there must be generous scope for changes in the method and intensity of cultivation, cropping and stocking; indeed, fuller cultivation would 'enhance rather than diminish the scenic effect'.[48]

In the event, development went much, much further than Dower and Hobhouse had foreseen. National Park status never resulted (as had been widely expected) in state ownership of land, but even if it had done, things would hardly have been different. It could not save, for example, Snowdonia from copper mining and a nuclear power station with its own irradiated cat in 1958, Pembroke from an oil terminal also in 1958, the North York Moors from sixty square miles of conifer and reseeded pasture, plus the early-warning installation at Fylingdales in 1960 and the immense Boulby potash mine, the Peak from extensive limestone quarrying, Dartmoor and Northumberland from military training or Exmoor the loss of 9,500 acres of heather moor (16 per cent of the park area) under the plough between 1958 and 1976. Maybe Scotland did not miss much.[49]

The 1949 Act also established the Nature Conservancy whose writ in this case did run in Scotland. Its vision in the early years was a holistic one, intending to use science as a 'biological service' to benefit both wildlife and land users. In this respect its best remembered achievement came through the Toxic Chemicals and Wildlife Section set up in 1960, uncovering the extent of pesticide pollution in the food chain through the study of peregrines and sparrowhawks. But its own prewar roots in the Society for the Promotion of Nature Reserves ensured that from the start it also had a strong site-based emphasis. The new body set about establishing a network of National Nature Reserves, a policy which at first alarmed Lord Salisbury, minister in the Conservative administration in 1955. He accused it of 'spattering' the whole country with 'living museums', and although he was mollified after an internal departmental investigation, it was a warning shot across the Conservancy's bows.[50] In fact, only where the Conservancy could buy the land itself, as at Beinn Eighe and Rum in Scotland and Moor House in the Pennines, did it prove possible beyond peradventure to put conservation and scientific research first. After 1955, due to a mixture of political pressure and need to minimise spending, most National Nature Reserves were declared without being purchased, after management agreements with the owners. Then, very often, as at Mar Lodge and Glen Feshie in the Cairngorms, erosion of the biological interest continued almost unchecked.

The Nature Conservancy and its successor, the Nature Conservancy Council, also developed a designated network of Sites of Special Scientific Interest, but they, too, provided unsystematic and weak defence. Over 3,000 had been selected in Britain by 1975, but at this stage the designation was little more than a label stuck on the ground which the farmer could disregard at will. 'The SSSI system was certainly extensive', comments W. M. Adams of this period. 'Unfortunately it was singularly ineffective.'[51]

In the three decades following the 1947 and 1949 Acts, the ecological and physical characters of the countryside were altered in a way unimaginable to the Scott Committee or John Dower, through the extraordinary chemical and mechanical revolution in farming described in Chapter 3. In reality, the ownership of rural land had now become very much more like the ownership of a bale of cotton or a chair than had the ownership of urban property. Whereas the developer in the city could not destroy a Georgian terrace without the approval of the local planning committee (something he might or might not find difficult), a farmer could grub up an ancient wood that had been a community asset from time immemorial with no hindrance whatever apart from appeals to his better nature. Such appeals were usually seen by the recipient to be inappropriate. The Country Landowners' Association and the National Farmers' Union replied in 1967 to criticism over encroachments on the Exmoor National Park heathlands with the observation that farmers resent any 'unreasonable directions . . . which might in any way restrict their right – which is also their duty – to expand their enterprises and to improve their productive capacity to a maximum.'[52] Yet the public became restive. By 1973 even an expert in planning law like D. A. Bigham, broadly sympathetic to the status quo in agricultural planning, was asking 'whether or not the removal of ancient (even, in some cases, literally prehistoric) hedgerows be allowed to continue at the rate of 5,000 miles per annum'.[53]

By the end of the 1970s, moreover, the balance of power between town and country was seen by many to have shifted slightly. There was much greater awareness of environmental issues, fed by the *Torrey Canyon* oil spillage, Rachel Carson's *Silent Spring* and the pesticide alarm at home. With greater affluence, there was much greater opportunity to explore the countryside, especially by car. Membership of leading environmental groups began to climb steeply. The National Trust had had only 8,000 members in 1946; by 1970 it had around 200,000 and by 1980 about a million. The RSPB had 5,900 members in 1945; by 1970, 67,000 and by 1980, 320,000. In 1964, the thirty-six county naturalist trusts then in existence had a membership of 17,700; by 1981 there were forty-three trusts with a membership of 140,000. By contrast, the Ramblers' Association could in 1981 only muster 32,000, which goes far to explain why in the next decade there was much more campaigning for nature conservation than for the ancient cause of access.[54]

It is an important question as to how far this increase of interest was shared equally between northern Britain and the south. Philip Lowe has shown that there was indeed a considerable disparity between the two. Membership of amenity societies thinned out in the north of England, and became sparser still in Scotland (see Figure 6.1).[55] The same was true of other environmental bodies. This has enabled the Highland opponents of, for example, the RSPB to claim that it is an alien body, reflecting only the views of the briefly visiting English to whom the hills and glens are a playground. Yet the RSPB can now point to a very substantial membership in the Highlands compared to most other voluntary bodies and, in general, environmental organisations have multiplied and strengthened, albeit at a lower level, in parallel to developments further south.

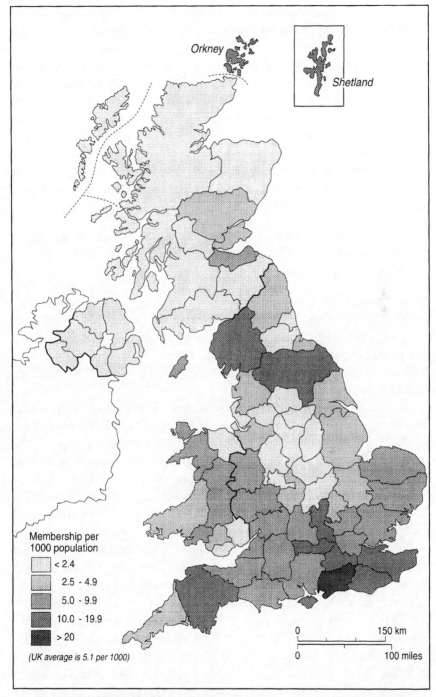

Figure 6.1 Geographical distribution of amenity society members, c. 1980.
Source: *Lowe and Goyder (1983).*

The multifaceted environmental lobby across the UK was already far bigger than any political party, but what effect did the new green monsters actually have on British politics? Battling with planning decisions over particular issues and at local level, wielding the twin weapons of 'public censure and delay', they were not to be trifled with.[56] From the 1970s onwards it became more difficult for authorities to intrude development into National Parks, build reservoirs in beautiful valleys or take more of the Cairngorms for ski development. Planners and ministers hated the skilful throwers of spanners into their works. Peter Walker, frustrated over a reservoir, deplored the tendency 'for the public to wish to retain all that exists and to oppose all that is new'.[57] His Labour equivalent, Anthony Crosland said much more fiercely:

> The conservationist lobby... is hostile to growth and indifferent to the needs of ordinary people. It has a manifestly class bias and reflects a set of middle and upper class value judgements.[58]

Yet the huge and rising membership of the environmental bodies, in contrast to that of the political parties, suggested an enviable base in popular support, and the RSPB in 1982, the largest of them, had only a quarter of its membership drawn from managerial and professional groups: the great bulk belonged to technical and clerical occupations, with 14 per cent from unskilled manual workers.[59]

Weight of numbers and subscriptions also made it easier for voluntary bodies to protect land by buying it, on the old model of the National Trust. By 1965 the RSPB had twenty-five reserves in Britain covering 2,800 hectares, but twenty years later it had ninety-five reserves covering 47,000 hectares. County naturalist trusts showed similar growth in their protective abilities, and covered another 55,000 hectares.[60]

However, the situation in respect to the influence of the environmental lobby on national legislation, notably where it touched upon the position of agriculture or the rights of landed property, was very different. Here the green monsters, for all their roaring, proved to have rubber teeth. In 1968 a new Countryside Act responded to public anxiety at the erosion of rural beauty. It allowed local authorities a more active role in rural planning, but left farming and forestry out of it again. It also gave Countryside Commissions in England and Scotland a general responsibility for safeguarding landscape without giving them appropriate powers to discharge it. A decade later Howard Newby could observe that despite local successes 'little impact seems to have been made upon the laissez-faire attitude of the farmer towards his own property'. That was the nub of the matter.[61]

Then, in 1977, the Nature Conservancy Council attempted to take the initiative in a report that emphasised the extent of the damage that agriculture was doing to biodiversity in the environment. It carefully avoided blaming the farmer by pointing out that it was inevitable under the current system of support payments, and argued that resources should be shifted to support the farmer in safeguarding land of high conservation interest under a national land-use strategy. It fell on deaf ministerial ears. The following year even the government's Countryside Review Committee, usually regarded as 'a talking shop on the English and Welsh countryside', became restive:

> For the first time in our history there is an increasing divergence between farming on the one hand and landscape and amenity conservation on the other; and the farmer, far from being accepted as a guardian of the countryside, is in danger of being regarded as its potential destroyer.[62]

Apart from wringing of hands, little was done. Meanwhile the erosion of semi-natural habitat proceeded unchecked. For example, by 1986, only 3 per cent of grassland in Cumbria remained unimproved, only 5 per cent of hay meadow in the Yorkshire Dales National Park was of SSSI quality, and 40 per cent of Highland birchwoods extant in 1940 had disappeared.

It was the 1981 Wildlife and Countryside Act that marked the high point of legislative achievement by the nature conservation lobby, but equally sharply illustrated its limitations. The Act came about because the British government needed to adapt UK legislation to meet the requirements of the EEC Birds Directive. It became a *cause célèbre* when an official in the NCC on his own initiative revealed in an article in the *New Scientist* the extent to which SSSIs under the old system were being damaged and lost. After fierce and prolonged debate, the Act introduced the present system of compensatory management agreements for SSSIs, and in effect increased at a stroke the proportion of land in Britain subject to conservation agreements by the NCC (and now by its successor bodies) from about 1 per cent (that is, the land which is National Nature Reserve) to about 10 per cent of the land surface.[63]

The Act was, however, still a defeat for those who, like Marion Shoard, argued that agriculture and forestry needed to be brought within the planning system like any other activity. Even the *Times* was sympathetic to this view. In a leader in 1985 it commented that 'many . . . do not believe that legislation short of that will ever be effective'.[64]

The upshot of the 1981 Act was to intensify the quarrel over the countryside. Financial guidelines for the 1981 Act, finally published at the end of 1984, produced compensatory management agreements that were likened to hot air balloons that could only be kept afloat by burning money.

Encouraging the osprey in Speyside:
Roy Dennis and a colleague in 1960, fixing an artificial
platform to the nest site at the first RSPB Loch Garten hide after the
tree had been vandalised.
Roy Dennis.

Disagreement over the terms of a particular settlement, followed by law
suits and arbitration, resulted, for example, in one wealthy Scottish business-
man and farmer securing well over £1,000,000 simply for not afforesting
land in Glen Lochay in Perthshire. There were bitter rows about Creag
Meagaidh in Inverness-shire, about the afforestation of the Flow Country
in Caithness and Sutherland, and about SSSIs on Islay. There were adjustments to the 1981 Act and to the fiscal regime of tax-breaks for forestry to
make them less of a public scandal, and a huge deflection of political and
bureaucratic effort into making the legislation workable. The level of
aggression and frustration on all sides led ultimately in 1991 to the disso-
lution of the NCC into different bodies in England, Wales and Scotland,
amid unavailing wails of protest from environmentalist organisations, but
the parameters of conflict in the countryside were only slightly changed.

What happened in the first five years to Scottish Natural Heritage,
successor to the NCC and the Countryside Commission for Scotland, is a
good illustration of this. Its founding statute mentioned the new concept of

'sustainability' for the first time in British legislation, and laid upon the organisation, in curious circumlocutary phrases, the duty of having regard to 'the desirability of securing that anything done, whether by SNH or any other person, in relation to the natural heritage of Scotland is undertaken in a manner which is sustainable'. When it attempted, however, to put this highly radical remit into operation by expressing its concerns over the wider environment – the 90 per cent of the land surface outside SSSIs – and by giving opinions on transport policy as exemplified by the proposals for a second Forth Road Bridge, it was (in effect) tartly reminded by the Scottish Office to keep in line or have its budget cut. Malcolm Rifkind, the Secretary of State who established SNH, had said at the Royal Society of Edinburgh prior to the organisation's launch that if it was not from time to time a thorn in his flesh it would not be doing its job. No one deemed it tactful at this juncture to remind his successor of these generous words.

Espousing a corporate ethos that deliberately sought consensus and partnership rather than confrontation, SNH scored one substantial success, in getting walkers, mountaineers and landowners to sit down together in an

Discouraging the osprey in Sutherland:
Charles St John planning to shoot a pair at their nest in 1848.

Access Forum that in due course paved the way for legislation that will enshrine a version of Bryce's Right to Roam. But in nature conservation it soon found itself in the same pot in which the NCC had stewed in the 1980s, declaring suites of SSSIs at the behest of a government fulfilling its European obligations in an atmosphere of non-consultation which was insisted upon by government, and incurring the wrath of crofters and farmers by whom it was portrayed as a quango of mad and secretive scientists out of touch with reality. Meanwhile it also failed to please the big environmental bodies, the RSPB and the World Wildlife Fund, who unsuccessfully took it to court for failing to draw the boundaries of designated land in the Cairngorms tightly enough.

In truth, the whole business over SSSIs was a sad and divisive distraction from what should have been the main business of all environmentalists, of bringing the maximum pressure to bear on reform of the Common Agricultural Policy, which affected not 10 per cent but (in Scotland) 74 per cent of the land surface, and whose effects were, by general consent, becoming more and more damaging. The Common Fisheries Policy bade fair to become even more ruinous to marine life in the firths and seas. Europe was nothing if not inconsistent, for there were initiatives from other parts of Brussels pointing in other directions. One was the programme to help farming in defined Environmentally Sensitive Areas, of which the Scottish Office made much less use than the English ministries. Another was the inauguration of a programme of Natura 2000 sites across Europe, which in the UK would have the effect of protecting more stringently the most important SSSIs but sadly and inevitably of reducing the level of protection on the others. Perhaps it was also true that the Countryside Alliance in 1997 should have taken its Land Rovers and Labradors to Brussels and Strasbourg rather than to London, but we are all slow to see where real sovereignty has gone since Mrs Thatcher signed the Single European Act.

The argument of this final chapter has been that the quarrel over the countryside is an argument over the limits and rights of property, beginning in the nineteenth century about access and widening in the twentieth to encompass also landscape protection and nature conservation. It can become so bitter because it goes to the root of the question – what does it mean to own and use land? For whom is the benefit and how is it shared? The argument of this book has been that use and delight are constant and often conflicting aspects of the human attitude to nature. I can see no end to the quarrel unless two conditions are met. Firstly, the countryside must recognise that the town has a perfectly legitimate interest in the countryside, encompassing a right of access (however defined) and a right to find

the countryside beautiful and full of diverse wildlife. It needs to be recognised that this land is indeed our land, and that owning landed property is
indeed not like owning a chair.

Secondly, the town – in this context both the wider public and the
narrower environmental movement – must respect that the countryman
needs to use the land, that the Common Agricultural Policy needs reform
to give the countryman sensible choice in the use of the land, and that if
need be the countryman needs to be paid above the rate of economic
return as reward for providing public benefits. There is certainly money
made out of rural recreation, conservation and tourism, and in general
terms it is vital support for a faltering rural economy, yet it seldom reaches
the pockets of those who own or farm the land. To divert even a little of the
immense sums that go into paying farmers to produce food in a Europe of
overproduction into paying them for truly being guardians of the countryside might seem like common sense applied to the common good, yet it
remains beyond the grasp of any British politician to bring it about. Such
changes, however, will demand more than political determination, important
though that is. They also demand a reduction in the sense of aggrieved
self-righteousness on all sides.

Yet the task of changing attitudes is going to be extremely formidable.
Twenty years ago the distinguished rural sociologist Howard Newby wrote
as follows:

> It is by no means clear how farmers would react to any system which
> threatened to control their freedom to do as they please with their own land.
> There is also little evidence to show that farmers welcome an opportunity
> to become glorified park wardens or landscape gardeners – nor, indeed
> have they the necessary skills to do so . . . farmers still remain suspicious
> of environmentalists [and] this suspicion runs very deep. It is not simply
> xenophobia, but a reluctance to admit any other 'proprietary interest',
> including that of the environmental lobby, into the control of private prop-
> erty. Any infringement of the owner's exclusive rights will be resisted, and
> for this reason, if no other, attempts to provide a planned resolution of the
> conflict between farming and the environment will prove difficult.[65]

These words are at least as true now as they were in 1979, but the
external pressures have moved somewhat. Firstly, farming is in a terrible
mess when a sheep sells for the price of a bottle of coke. Secondly, the use
of the countryside from visitors grows and grows, as do the numbers
joining environmental bodies. Thirdly, government itself is beginning to
bend, at least in Scotland, more towards an older view of landed property.
Thus a Scottish Office paper of 1998, *People and Nature*, discussing the

future of SSSIs in Scotland, used words very different from the sentiments of the CLA and NFU on Exmoor in 1967:

> Land is a resource which has a value not only for the individual owners and occupiers, but also for the wider community who can derive benefits or incur costs according to how particular areas of land are managed . . . The arrangements for the management of land of high nature conservation value should reflect this. Therefore, the wider local and national communities have a legitimate interest in the management of such land, and it is unreasonable for individual landowners or occupiers to expect to be able to manage such land purely in terms of their own private interest.[66]

Here is the same recognition of the public nature of private property, of the right of the town to the country, as we found implicit in the words of William Wordsworth, Thomas Bewick, John Stuart Mill, John Ruskin, Canon Rawnsley and Professor Joad. It does not solve the vexed problem of the limits of public interest, or how and by whom that interest is identified and represented. By focusing on special sites of high nature conservation interest, government still runs the concomitant danger of suggesting that the public can derive no benefits and have no claim on land outside them. But it seems to acknowledge that in this small, old country, where nothing is wilderness but people are long practised in agriculture, use of nature and delight in it need to be reunited, seamlessly and quickly. If we ignore the claims of the public on the wider countryside – their right to find it beautiful and full of natural life – and simultaneously ignore the economic plight of farmers who will be forced by circumstance either to abandon the land or to use it so intensively that it becomes a greater desert, there will be little delight left for our children's children. Then the quarrel over the countryside will prove to have been sterile, futile, endless and sad.

NOTES

NOTES TO INTRODUCTION

1. This is not to overlook or devalue the work of T. M. Devine, *The Transformation of Rural Scotland: Social Change and the Agrarian Economy, 1660–1815* (Edinburgh, 1994), R. A. Dodgshon, *Land and Society in Early Scotland* (Oxford, 1981) and I. D. Whyte, *Agriculture and Society in Seventeenth Century Scotland* (Edinburgh, 1979). What is truly extraordinary is the absence of Scottish agricultural history in the nineteenth and twentieth centuries.
2. For astute detection of 'neo-Whiggish inclinations' in the environmental history of Richard Grove, Keith Thomas and myself, see J. M. Mackenzie, *Empires of Nature and the Nature of Empires* (Edinburgh, 1997), p. 19.
3. J. S. Mill, *Principles of Political Economy*, ed. W. J. Ashley, 1909, book 4, chapter 6.

NOTES TO CHAPTER 1

1. M. W. Holdgate, 'Standards, sustainability and integrated land use', *Macaulay Land Use Research Institute 10th Anniversary Lectures* (Aberdeen, 1997), p. 13.
2. T. Burnet, *The Sacred Theory of the Earth* (London edn, 1965), especially book 1, chapters 4 and 5. For his influence and ideas, see J. Wyatt, *Wordsworth and the Geologists* (Cambridge, 1995), pp. 47–8; M. H. Nicholson, *Mountain Gloom and Mountain Glory: The Development of the Aesthetics of the Infinite* (New York, 1959); S. J. Gould, *Ever Since Darwin* (Harmondsworth edn, 1980), pp. 141–6; S. Schama, *Landscape and Memory* (London, 1995), p. 451; K. Olwig, *Nature's Ideological Landscape* (London, 1984), pp. 32–3.
3. D. Defoe, *A Tour through the Whole Island of Great Britain*, eds G. D. H. Cole and D. C. Browning (London, 1974), p. 271.
4. E. Burt, *Letters from the North of Scotland to his Friend in London*, ed. R. Jamieson (Edinburgh, 1876), p. 32.
5. S. Johnson, *A Journey to the Western Islands of Scotland in 1773* (London, 1876), p. 32.
6. R. Noyes, *Wordsworth and the Art of Landscape* (New York, 1973), p. 68.
7. A. Mitchell (ed.), *Geographical Collections Relating to Scotland, Made by Walter Macfarlane* (Scottish History Society, Edinburgh, 1906), vol. 2, pp. 109–10.
8. J. C. Stone, *The Pont Manuscript Maps of Scotland: Sixteenth-century Origins of a Blaeu Atlas* (Tring, 1989), pp. 16–38.
9. *Geographical Collections*, vol. 2, p. 540.

10. Burnet, *Sacred Theory*, pp. 109–11.

11. Alexander Pennecuik, *Description of Tweeddale* (1715) in his *Works* (Leith, 1815), p. 44.

12. New Register House, Edinburgh, Parish Register of St Andrews, OPR 453/9, p. 139.

13. G. Murphy (ed.), *Early Irish Lyrics: Eighth to Twelfth Centuries* (Oxford, 1956), p. 161. See also D. Thomson, *An Introduction to Gaelic Poetry* (London, 1974) and J. Hunter, *On the Other Side of Sorrow: Nature and People in the Scottish Highlands* (Edinburgh, 1995).

14. Cited in D. S. Thomson (ed.), *The Companion to Gaelic Scotland* (Oxford, 1994), p. 212.

15. Thomson (ed.), *Companion*, pp. 212–13.

16. A. Macleod (ed. and trans.), *The Songs of Duncan Ban Macintyre* (Scottish Gaelic Texts Society, Edinburgh, 1952), pp. 165–9.

17. *Geographical Collections*, vol. 2, p. 288.

18. *Geographical Collections*, vol. 3, p. 144; H. J. Cook, 'Physicians and natural history', in N. Jardine, J. A. Secord and E. C. Spary (eds), *Cultures of Natural History* (Cambridge, 1996), p. 99.

19. Sir J. Sinclair (ed.), *The Statistical Account of Scotland 1791–1799* (eds D. J. Withrington and I. R. Grant, Wakefield), vol. 10, *Fife* (1978), p. 467. The parish concerned was Kilmany in Fife.

20. T. Carlyle, *Works* (London, 1829), vol. 2, pp. 59–60.

21. *Geographical Collections*, vol. 1, p. 248; R. Rennie, *Essays on the Natural History and Origins of Peat Moss* (Edinburgh, 1807), p. 6; A. Steele, *The Natural and Agricultural History of Peat-moss or Turf-bog* (Edinburgh, 1826).

22. W. Aiton, *A Treatise on the Origin, Qualities and Cultivation of MossEarth, with Directions for Converting it into Manure* (Ayr, 1811), pp. 341–2.

23. Noyes, *Art of Landscape*, pp. 5–8.

24. S. Monk, *The Sublime: A Study of Critical Theories in Eighteenth Century England* (New York, 1935), pp. 84–125.

25. Schama, *Landscape and Memory*, p. 450.

26. W. Gilpin, *Observations Relative Chiefly to Picturesque Beauty, made in the Year 1776 on Several Parts of Great Britain, particularly in the High-lands of Scotland* (London, 1792), vol. 1, pp. 119–23.

27. Ibid., pp. 124–5.

28. Monk, *Sublime*, pp. 227–32.

29. W. Wordsworth, *Prose Works*, ed. A. B. Gosart (London, 1876), vol. 3, p. 244.

30. *Lord of the Isles*, Canto 4, lines 3–4, 18–27.

31. P. Womack, *Improvement and Romance: Constructing the Myth of the Highlands* (London, 1985), p. 172.

32. J. Bate, *Romantic Ecology: Wordsworth and the Environmental Tradition* (London, 1991).

33. Noyes, *Art of Landscape*, pp. 96–111.

34. H. Taylor, *A Claim on the Countryside: A History of the British Outdoor Movement* (Edinburgh, 1997), p. 20.

35. J. Garritt, 'Politics, knowledge, action: the local implementation of the Convention on Biological Diversity', unpublished conference paper 1998, cited with permission.

36. L. Koerner, 'Carl Linnaeus in his time and place', in Jardine, Secord and Spary, *Cultures of Natural History*, p. 157; K. Thomas, *Man and the Natural World: Changing Attitudes in England, 1500–1800* (London, 1983); D. E. Allen, *The Naturalist in Britain: A Social History* (London, 1976), p. 197; B. Harrison, 'Animals and the state in nineteenth-century England', *English Historical Review*, vol. 349 (1973), pp. 786–820.

37. R. Lambert, 'From exploitation to extinction to environmental icon: our images of the Great Auk', in R. Lambert (ed.), *Species History in Scotland: Introductions and Extinctions since the Ice Age* (Edinburgh, 1998), p. 27.

38. In a broadcast on BBC Radio 4, 19 November 1998.

39. Quoted in J. Sheail, *Nature in Trust: The History of Nature Conservation in Britain* (Glasgow, 1976), p. 5.

40. F. O. Morris, *A History of British Butterflies* (6th edn, London, 1891), p. 116.

41. C. Nairne, 'Perthshire', in G. Scott Moncrieff (ed.), *Scottish Country* (n.p., 1935), pp. 242–3.

42. Allen, *Naturalist*, pp. 197–8; Sheail, *Nature in Trust*, pp. 23–4.

43. Sheail, *Nature in Trust*, p. 24.

44. Ibid., p. 35.

45. J. Sheail, *Seventy-five Years in Ecology: the British Ecological Society* (Oxford, 1987), p. 138.

46. W. M. Adams, 'Rationalization and conservation: ecology and the management of nature in the United Kingdom', *Transactions of the Institute of British Geographers*, vol. 22 (1997), pp. 277–91.

47. *Geographical Collections*, vol. 2, pp. 24, 181; vol. 3, p. 314.

48. C. St John, *Short Sketches of the Wild Sports and Natural History of the Highlands* (London, 1847), pp. 227–8.

49. J. Sheail, *Nature in Trust*, pp. 37–9; F. Fraser Darling and J. M. Boyd, *The Highlands and Islands* (London, 1964), pp. 253–4; R. J. Berry and J. L. Johnston, *The Natural History of Shetland* (London, 1980), pp. 113–16; *Scottish Natural Heritage, The Natural Heritage of Scotland: An Overview* (Perth, 1995), pp. 145, 152.

50. Quoted in J. W. Kempster, *Our Rivers* (Oxford, 1948).

51. *The Scotsman*, 15 and 16 January 1999.

52. Fergus Ewing MP, in a broadcast on BBC Radio 4, 19 February, 1999.

53. Lord Sewel, ibid.

54. R. Mabey, *The Common Ground* (London, 1980), p. 25.

55. J. Ruskin, *Works* (eds E. T. Cook and A. Wedderburn, London, 1905), vol. 17, p. lxxxix.

NOTES TO CHAPTER 2

1. H. Miles and B. Jackman, *The Great Wood of Caledon* (Lanark, 1991), pp. 11–12.

2. R. Tipping, 'The form and fate of Scotland's woodlands', *Proceedings of the Society of Antiquaries of Scotland*, vol. 124 (1994), pp. 1–54.

3. J. Ritchie, *The Influence of Man on Animal Life in Scotland* (Cambridge, 1920); A. C. Kitchener, 'Extinctions, introductions and colonisations of Scottish mammals and birds since the last Ice Age', in R. A. Lambert (ed.),

Species History in Scotland: Introductions and Extinctions since the Ice Age (Edinburgh, 1998), pp. 63–92.

4. Tipping, 'Form and fate'.

5. D. Breeze, 'The Great Myth of Caledon', in T. C. Smout (ed.), *Scottish Woodland History* (Edinburgh, 1997), pp. 47–51.

6. See, in general, Smout (ed.), *Scottish Woodland History*, esp. chs 1, 7 and 9.

7. F. Fraser Darling, *Natural History in the Highlands and Islands* (London, 1947), p. 59.

8. *Green Party Manifesto for the Highlands*, 1990. Bernard Planterose was leader of Reforesting Scotland when he penned the manifesto.

9. Scottish Council (Development and Industry), *National Resources in Scotland: Symposium at the Royal Society of Edinburgh* (Edinburgh, 1961), p. 338.

10. H. A. Maxwell, 'Coniferous plantations', in J. Tivy (ed.), *The Organic Resources of Scotland: Their Nature and Evaluation* (Edinburgh, 1973), p. 182.

11. Breeze, 'The Great Myth'; A. Birley, *Septimus Severus: The African Emperor* (London, 1971), pp. 255–7.

12. D. Breeze, *Roman Scotland* (Edinburgh, 1996); J. H. Dickson, 'Scottish woodlands: their ancient past and precarious future', *Scottish Forestry*, vol. 47 (1993), pp. 73–8.

13. R. J. Mercer et al., 'The Early Bronze Age cairn at Sketewan, Balnaguard, Perth and Kinross', *Proceedings of the Society of Antiquaries of Scotland*, vol. 127 (1997), pp. 281–338; R. Tipping, A. Davies and E. Tisdall, *West Affric Forest Restoration Initiative, Draft Interim Report, Year 2* (National Trust for Scotland, 1999), pp. 17, 49.

14. Breeze, *Roman Scotland*, p. 97.

15. P. H. Brown (ed.), *Scotland before 1800 from Contemporary Documents* (Edinburgh, 1893), pp. 80–1.

16. R. Sibbald, *Scotia Illustrata* (Edinburgh, 1684).

17. G. Chalmers, *Caledonia, or a Historical and Topographical Account of North Britain from the Most Ancient to the Present Times* (new edn, Paisley, 1887).

18. L. Shaw, *The History of the Province of Moray* (edn Glasgow, 1882), vol. 3, p. 11; T. Pennant, *Tour in Scotland in 1769* (edn Warrington, 1774), pp. 93, 109, 115, 212–13.

19. J. E. Bowman, *The Highlands and Islands, A Nineteenth-Century Tour* (Gloucester, 1986), pp. 161, 163.

20. J. S. Stuart and C. E. Stuart, *Lays of the Deer Forest* (Edinburgh, 1848), vol. 2, pp. 256–7.

21. Ibid., pp. 220–1.

22. J. Radkau, 'The wordy worship of nature and the tacit feeling for nature in the history of German forestry', in M. Teich, R. Porter, B. Gustafsson (eds), *Nature and Society in Historical Context* (Cambridge, 1997), pp. 228–39; S. Schama, *Landscape and Memory* (London, 1995), p. 107.

23. W. F. Skene, *Celtic Scotland* (Edinburgh, 1871), vol. 1, pp. 84–6.

24. D. Nairne, 'Notes on Highland woods, ancient and modern', *Transactions of the Gaelic Society of Inverness*, vol. 17 (1892), pp. 170–221.

25. G. J. Walker and K. J. Kirby, *Inventories of Ancient, Long-established and Semi-natural Woodland for Scotland*, Nature Conservancy Council Research and Survey in Nature Conservation, no. 22 (1989); Nairne, '*Notes*', p. 191.

26. J. C. Stone, *The Pont Manuscript Maps of Scotland: Sixteenth-century Origins of a Blaeu Atlas* (Tring, 1989); A. Mitchell (ed.), *Macfarlane's Geographical Collections Relating to Scotland, Made by Walter Macfarlane* (Scottish History Society, Edinburgh, 1906), 3 vols.

27. Kitchener, 'Extinctions'.

28. A. J. L. Winchester, *Landscape and Society in Medieval Cumbria* (Edinburgh, 1987), pp. 100–7.

29. S. Barber, 'The history of the Coniston woodlands, Cumbria, UK', in K. J. Kirby and C. Watkins (eds), *The Ecological History of European Forests* (Wallingford, 1998), pp. 167–83.

30. R. Gulliver, 'What were woods like in the seventeenth century? Examples from the Helmsley Estate, North-east Yorkshire', in Kirby and Watkins (eds), *Ecological History*, pp. 135–54.

31. A. Fleming, 'Towards a history of wood pasture in Swaledale (North Yorkshire)', *Landscape History*, vol. 19 (1997), pp. 57–74.

32. Details of Scottish woodland management in this and the following paragraphs are drawn from research in progress; for a preliminary report, see C. Smout and F. Watson, 'Exploiting semi-natural woods, 1600–1800', in Smout (ed.), *Scottish Woodland History*, pp. 86–98. See also J. M. Lindsay, 'The Use of Woodland in Argyllshire and Perthshire between 1650 and 1850', unpublished University of Edinburgh PhD thesis, 1974.

33. A. Thomson, 'The Scottish Timber Trade, 1680–1800', unpublished University of St Andrews PhD thesis, 1990; A. Lillehammer, 'The Scottish–Norwegian timber trade in the Stavanger area in the sixteenth and seventeenth centuries', in T. C. Smout (ed.), *Scotland and Europe, 1200–1850* (Edinburgh, 1986), pp. 97–111.

34. A. Smith, *An Inquiry into the Nature and Causes of the Wealth of Nations*, eds R. H. Campbell and A. S. Skinner (Oxford, 1976), vol. 1, p. 183.

35. *Geographical Collections*, vol. 2, p. 3.

36. Ibid., vol. 2, p. 70.

37. Ibid., vol. 3, p. 242.

38. R. Callander, *History in Birse* (Finzean, 1981–5), nos 1–4.

39. F. Watson, 'Rights and responsibilities: wood-management as seen through baron court records', in Smout (ed.), *Scottish Woodland History*, pp. 101–14; F. Watson, 'Need versus greed? Attitudes to woodland management on a central Scottish Highland estate, 1630–1740', in C. Watkins (ed.), *European Woods and Forests: Studies in Cultural History* (Wallingford, 1998), pp. 135–56.

40. C. Innes (ed.), *The Black Book of Taymouth* (Ballantyne Society, Edinburgh, 1855), pp. 352–9.

41. Mitchell (ed.), *Geographical Collections*, vol. 2, pp. 272–3.

42. R. A. Dodgshon, *Land and Society in Early Scotland* (Oxford, 1981), pp. 290–2.

43. Smith, *Wealth of Nations*, vol. 1, p. 183.

44. H. H. Lamb, 'Climate and landscape in the British Isles' in S. R. J. Woodell (ed.), *The English Landscape, Past, Present and Future* (Oxford, 1985), p. 153; J. Grove, *The Little Ice Age* (London, 1988).

45. Lamb, 'Climate and Landscape', p. 155.

46. Stone, *Pont Manuscript Maps*; Mitchell (ed.), *Geographical Collections*, vol. 2, p. 165.

47. G. Mackenzie, Earl of Cromartie, 'An account of the mosses in Scotland', *Philosophical Transactions of the Royal Society*, vol. 27 (1710–12), pp. 296–301.

48. F. Watson, 'Sustaining a myth: the Irish in the West Highlands', *Scottish Woodland History Discussion Group Notes*, vol. 2 (1997), pp. 7–9.

49. National Library of Scotland MS 1359.100.

50. J. Henderson, *General View of the Agriculture of the County of Sutherland* (London, 1812), pp. 83–6, 1–6, 176.

51. J. M. Lindsay, 'Charcoal iron smelting and its fuel supply; the example of Lorn Furnace, Argyllshire, 1753–1876', *Journal of Historical Geography*, vol. 1 (1975), pp. 283–98.

52. Lindsay, thesis, especially chapters 8 and 9.

53. M. Jones, 'The rise, decline and extinction of spring wood management in south-west Yorkshire', in Watkins (ed.), *European Woods*, pp. 55–71.

54. R. Monteath, *Miscellaneous Reports on Woods and Plantations* (Dundee, 1827), p. 121.

55. C. Dingwall, 'Coppice management in Highland Perthshire', in Smout (ed.), *Scottish Woodland History*, pp. 162–75, citation on p. 171.

56. C. Smout, 'Cutting into the pine: Loch Arkaig and Rothiemurchus in the eighteenth century' and B. M. S. Dunlop, 'The woods of Strathspey in the nineteenth and twentieth centuries', in Smout (ed.), *Scottish Woodland History*, pp. 115–25, 176–89; C. Smout, 'The history of the Rothiemurchus woodlands', in T. C. Smout and R. Lambert (eds), *Rothiemurchus: Nature and People on a Highland Estate, 1500–2000* (Edinburgh, 1999).

57. D. Nairne, 'Notes on Highland woods', p. 220.

58. A. Watson, 'Eighteenth-century deer numbers and pine regeneration near Braemar, Scotland', *Biological Conservation*, vol. 25 (1983), pp. 289–305.

59. Dunlop, 'Woods of Strathspey', pp. 180–8.

60. K. Thomas, *Man and the Natural World: Changing Attitudes in England, 1500–1800* (London, 1984), pp. 192–7.

61. W. Daniels, *The Life of Ailred of Rievaulx*, trans. F. M. Powicke (London, 1950), p. 12.

62. Brown (ed.), *Scotland before 1800 from Contemporary Documents*, p. 142; Mitchell (ed.), *Geographical Collections*, vol. 2, p. 544.

63. J. Veitch, *Feeling for Nature in Scottish Poetry* (Edinburgh, 1887), vol. 2, pp. 11–13.

64. H. Cheape, 'Woodlands on the Clanranald estates', in T.C. Smout (ed.), *Scotland since Prehistory: Natural Change and Human Impact* (Aberdeen, 1993), p. 60.

65. T. C. Smout, 'Trees as historic landscapes: Wallace's oak to Reforesting Scotland', *Scottish Forestry*, vol. 48 (1994), p. 246.

66. Ibid., pp. 247–8.

67. Cited in Veitch, *Feeling for Nature*, vol. 2, pp. 152–3.

68. H. Cockburn, *Circuit Journeys* (Edinburgh, 1888), pp. 39–40.

69. J. Sheail, *Rural Conservation in Interwar Britain* (Oxford, 1981), pp. 172–5.

70. G. Ryle, *Forest Service: The First Forty-five Years of the Forestry Commission of Great Britain* (Newton Abbot, 1969), p. 259.
71. W. H. Murray, *Highland Landscape: A Survey* (Edinburgh, 1962).
72. Sheail, *Rural Conservation*, pp. 172–86.
73. Scottish Record Office: Forestry Commission 9/1. Commission meeting 19 January 1944.
74. J. Tsouvalis-Gerber, 'Making the invisible visible: ancient woodlands, British forest policy and the social construction of reality', in Watkins (ed.), *European Woods*, pp. 215–29.

NOTES TO CHAPTER 3

1. R. Mercer and R. Tipping, 'The prehistory of soil erosion in the northern and eastern Cheviot hills, Anglo-Scottish Borders', in S. Foster and T. C. Smout (eds), *The History of Soils and Field Systems* (Aberdeen, 1994), pp. 14–16.
2. E. I. Newman and P. D. A. Harvey, 'Did soil fertility decline in medieval English farms? Evidence from Cuxham, Oxfordshire, 1320–1340', *Agricultural History Review*, vol. 45 (1997), pp. 119–36; J. Pretty, 'Sustainable agriculture in the Middle Ages: the English manor', *Agricultural History Review*, vol. 38 (1990), pp. 1–19. More widely, see G. Clark, 'The economics of exhaustion, the Postan thesis, and the Agricultural Revolution', *Journal of Economic History*, vol. 52 (1992), pp. 61–84.
3. R. S. Shiel, 'Improving soil productivity in the pre-fertiliser era', in B. M. S. Campbell and M. Overton (eds), *Land, Labour and Livestock: Historical Studies in European Agricultural Productivity* (Manchester, 1991), p. 62.
4. Messrs Rennie, Brown and Shirreff, *General View of the Agriculture of the West Riding of Yorkshire* (London, 1794), p. 26; J. Shaw, 'Manuring and fertilising the Scottish lowlands', in Foster and Smout (eds), *Soils and Field Systems*, pp. 110–18.
5. T. Kjærgaard, *The Danish Revolution, 1500–1800; An Ecohistorical Interpretation* (Cambridge, 1994), p. 49.
6. M. M. Postan, *Essays in Medieval Agriculture and General Problems of the Medieval Economy* (Cambridge, 1973), pp. 3–27, 150–85.
7. Kjærgaard, *Danish Revolution*, especially chapter 2.
8. A. Mitchell (ed.), *Geographical Collections Relating to Scotland made by Walter Macfarlane* (Scottish History Society, Edinburgh, 1906), vol. 2, p. 140.
9. R. E. Tyson, 'Contrasting régimes; population growth in Ireland and Scotland during the eighteenth century', in S. J. Connolly et al. (eds), *Conflict Identity and Economic Development, Ireland and Scotland 1600–1939* (Preston, 1995), p. 67; T. C. Smout, N. C. Landsman and T. M. Devine, 'Scottish emigration in the seventeenth and eighteenth centuries', in N. Canny (ed.), *Europeans on the Move* (Oxford, 1994), pp. 77–90; A. Gibson and T. C. Smout, 'Scottish food and Scottish history, 1500–1800', in R. A. Houston and I. D. Whyte (eds), *Scottish Society 1500–1800* (Cambridge, 1989), pp. 59–84; A. J. S. Gibson and T. C. Smout, *Prices, Food and Wages in Scotland, 1550–1780* (Cambridge, 1995), especially chapters 7 and 9.

10. R. E. Prothero, *English Farming Past and Present* (London, 1917), pp. 146–7; J. Sheail, *Regional Distribution of Wealth in England as Indicated in the 1524–5 Lay Subsidy Returns*, List and Index Society Special Series, vol. 28 (1998), p. xii.

11. Quoted in L. Colley, *Britons: Forging the Nation 1707–1837* (Yale, 1992), p. 16.

12. T. C. Smout, *Scottish Trade on the Eve of Union, 1660–1707* (Edinburgh, 1963), p. 207; D. Ure, *General View of the Agriculture of the County of Roxburgh* (London, 1794), p. 26.

13. T. M. Devine, *The Transformation of Rural Scotland: Social Change and the Agrarian Economy, 1660–1815* (Edinburgh, 1994), p. 56.

14. J. Robertson, *General View of the Agriculture in the Southern Districts of the County of Perth* (Edinburgh, 1794), p. 26.

15. J. Bailey and G. Culley, *General View of the Agriculture of the County of Northumberland* (London, 1794), p. 44; Mr Tuke, *General View of the Agriculture of the North Riding of Yorkshire* (London, 1794), p. 52.

16. National Library of Scotland MS 33.5.16, Sir Robert Sibbald, 'Discourse anant the Improvements may be made in Scotland for advancing the Wealth of the Kingdom, 1698'.

17. Quoted in R. J. Brien, *The Shaping of Scotland: Eighteenth Century Patterns of Land Use and Settlement* (Aberdeen, 1989), p. 43.

18. D. A. Davidson and I. A. Simpson, 'Soils and landscape history: case studies from the Northern Isles of Scotland', in Foster and Smout (eds), *Soils and Field Systems*, pp. 66–74.

19. J. Donaldson, *Husbandry Anatomised, or an Enquiry into the present Manner of Tilling and Manuring the Ground*, quoted in D. Woodward, '"Gooding the Earth": Manuring Practices in Britain, 1500–1800', in Foster and Smout (eds), *Soils and Field Systems*, p. 103.

20. Tuke, *North Riding*, p. 51; Bailey and Culley, *Northumberland*, p. 45; W. Fullarton, *General View of the Agriculture of the County of Ayr* (London, 1794), p. 52; J. Naismith, *General View of the Agriculture of the County of Clydesdale* (London, 1794), p. 62.

21. D. M. Henderson and J. H. Dickson (eds), *A Naturalist in the Highlands: James Robertson, His Life and Travels in Scotland, 1767–1771* (Edinburgh, 1994), p. 29; G. S. Keith, *General View of the Agriculture of Aberdeenshire* (London, 1811), p. 432; A. Wight, *Present State of Husbandry in Scotland* (Edinburgh, 1778–84), vol. 3, pp. 599–600.

22. Ibid.

23. A. Fenton, *The Shape of the Past, 2: Essays in Scottish Ethnology* (Edinburgh, 1986), p. 59.

24. Tuke, *North Riding*, p. 51.

25. R. A. Dodgshon, 'Land improvement in Scottish farming: marl and lime in Roxburghshire and Berwickshire in the eighteenth century', *Agricultural History Review*, vol. 26 (1978), pp. 1–14.

26. I. Leatham, *General View of the Agriculture of the East Riding of Yorkshire* (London, 1794), p. 55.

27. Fenton, *Shape of the Past*, p. 66.

28. J. Sinclair (ed.), *The Statistical Account of Scotland*, vol. 2, *The Lothians* (D.

Withrington and I. Grant edn, 1975), p. 225.

29. G. Robertson, *General View of the County of Midlothian* (Edinburgh, 1793), pp. 48–9.
30. Leatham, *East Riding*, pp. 53–4.
31. Sinclair, *Statistical Account*, vol. 10, *Fife* (edn. 1978), pp. 308–9.
32. Rennie, Brown and Shireff, *West Riding*, p. 30.
33. Quote in Fenton, *Shape of the Past, 2*, p. 85.
34. R. A. Dodgshon and E. G. Olsson, 'Productivity and nutrient use in eighteenth-century Scottish Highland townships', *Geografiska Annaler*, vol. 70B (1988), pp. 39–51; R. A. Dodgshon, 'Strategies of farming in the western Highlands and Islands prior to crofting and the clearances', *Economic History Review*, vol. 46 (1993), pp. 679–701; R. A. Dodgshon, 'Budgeting for survival: nutrient flow and traditional Highland farming', in Foster and Smout (eds), *Soils and Field Systems*, pp. 83–93.
35. E. Boserup, *The Conditions of Agricultural Growth: The Economics of Agrarian Change under Population Pressure* (London, 1965); E. Boserup, *Population and Technology* (Oxford, 1981).
36. B. Falkner, *The Muck Manual: A Practical Treatise on the Nature and Value of Manure* (London, 1843).
37. Shiel, 'Soil productivity', p. 67.
38. H. Stephens, *Manual of Practical Draining* (Edinburgh, 1846), p. 11.
39. T. H. Nelson, *The Birds of Yorkshire: A Historical Account of the Avifauna of the County* (London, 1907), vol. 1, p. 216; E. V. Baxter and L. J. Rintoul, *The Birds of Scotland: Their History, Distribution and Migration* (Edinburgh, 1953), vol. 1, p. 38.
40. J. Sheail, 'Elements of sustainable agriculture: the UK experience, 1840–1940', *Agricultural History Review*, vol. 43 (1995), pp. 178–92.
41. *The Country Gentleman's Catalogue* (London, 1894), pp. 190–2.
42. K. Blaxter and N. Robertson, *From Dearth to Plenty: The Modern Revolution in Food Production* (Cambridge, 1995), pp. 28–33.
43. H. Newby, *Green and Pleasant Land? Social Change in Rural England* (Harmondsworth, 1980), chapter 4.
44. Blaxter and Robertson, *Dearth to Plenty*, pp. 27, 56; R. J. O'Connor and M. Shrubb, *Farming and Birds* (Cambridge, 1986), pp. 82–3.
45. Blaxter and Robertson, *Dearth to Plenty*, p. 80.
46. O'Connor and Shrubb, *Farming and Birds*, p. 82.
47. Ibid., pp. 191, 196; K. Mellanby, *Farming and Wildlife* (London, 1981), p. 106.
48. D. A. Davidson and T. C. Smout, 'Soil change in Scotland', in A. G. Taylor, J. E. Gordon, and M. B. Usher (eds), *Soils, Sustainability and the Natural Heritage* (Edinburgh, 1996), pp. 44–54.
49. L. B. Powell, 'Deteriorating soil', in E. Goldsmith (ed.), *Can Britain Survive?* (London, 1971), p. 71.
50. Mellanby, *Farming and Wildlife*, pp. 57–63.
51. *The Natural Heritage of Scotland, an Overview* (Scottish Natural Heritage, Edinburgh, 1991), p. 84; E. C. Mackey, M. C. Shewry and G. J. Tudor, *Land Cover Change: Scotland from the 1940s to the 1980s* (Edinburgh, 1998).
52. J. Miles, 'The soil resource and problems today', in Foster and Smout (eds), *Soils and Field Systems*, pp. 146–7.

53. See Taylor, Gordon and Usher (eds), *Soils, Sustainability*, p. xv.

54. Mellanby, *Farming and Wildlife*, pp. 64–5.

55. Davidson and Smout, 'Soil change in Scotland', pp. 45–7.

56. M. B. Usher, 'The soil ecosystem and sustainability', in Taylor, Gordon and Usher (eds), *Soils, Sustainability*, pp. 22–43.

57. N. W. Moore, *The Bird of Time: The Science and Politics of Nature Conservation* (Cambridge, 1987), pp. 142–58.

58. P. Bassett, *A List of the Historical Records of the Royal Society for the Protection of Birds* (Birmingham, 1980), p. v.

59. Moore, *Bird of Time*, pp. 149, 159–62.

60. J. Sheail, *Pesticides and Nature Conservation: The British Experience, 1950–1975* (Oxford, 1985).

61. O'Connor and Shrubb, *Farming and Birds*, pp. 201–5; Moore, *Bird of Time*, p. 174.

62. O'Connor and Shrubb, *Farming and Birds*, pp. 196–215. See also J. H. Marchant et al., *Population Trends in British Breeding Birds* (Thetford, 1990); *BTO News*, 207 (1966).

63. See *Fife Bird Report*, 1995–8; A. M. Smout, *The Birds of Fife* (Edinburgh, 1986) and personal observations.

64. M. Shrubb, letter in *British Birds*, vol. 91 (1998), p. 332; J. Ritchie, *The Influence of Man on Animal Life in Scotland* (Cambridge, 1920), p. 179; Isle of Wight archives, churchwarden's account of the Parish of Godshill, 1815–1820. I am indebted to Richard Smout, the archivist, for this record.

65. Marchant et al., *Population Trends*, pp. 217–18.

66. O. E. Prys-Jones and S. A. Corbet, *Bumblebees* (Slough edn, 1991), p. 89.

67. Mellanby, *Farming and Wildlife*, p. 41.

68. A. G. Bradley, *When Squires and Farmers Thrived* (London, 1927), p. 77.

NOTES TO CHAPTER 4

1. J. M. Hunter, *Land into Landscape* (Harlow, 1986), p. 10.

2. D. Kinnersley, *Troubled Water: Rivers, Politics and Pollution* (London, 1988), p. 14; D. J. Gilvear and S. J. Winterbottom, 'Changes in channel morphology, floodplain land use and flood damage on the rivers Tay and Tummell over the last 250 years: implications for floodplain management', in R. G. Bailey, P. V. José and B. R. Sherwood (eds), *United Kingdom Floodplains* (London, 1998), pp. 92–115.

3. Gilvear and Winterbottom, 'Changes in channel morphology', p. 95.

4. P. H. Brown, *Early Travellers in Scotland* (Edinburgh, 1891), pp. 266–7.

5. W. Smith (ed.), *Old Yorkshire* (London, 1883), pp. 48, 60–71; T. H. Nelson, *The Birds of Yorkshire: A Historical Account of the Avifauna of the County* (London, 1907), vol. 2, p. 438.

6. J. A. Sheppard, *The Draining of the Hull Valley* (East Yorkshire Local History Society, 1958), p. 6.

7. J. A. Sheppard, *Draining of the Marshlands of South Holderness* (East Yorkshire Local History Society, 1966), pp. 19–20.

8. A. Coney, 'Fish, fowl and fen: landscape economy in seventeenth-century Martin Mere', *Landscape History*, vol. 14 (1992), pp. 51–64.

9. E. V. Baxter and L. J. Rintoul, *The Birds of Scotland; Their History, Distribution and Migration* (Edinburgh, 1953), vol. 1, pp. 350–1; C. Hough, 'The trumpeters of Bemersyde', *Scottish Placename News*, no. 5 (1998), p. 3.

10. Nelson, *Birds of Yorkshire*, vol. 2, p. 399.

11. Sheppard, *Draining of the Marshlands*, p. 11; Smith, *Old Yorkshire*, pp. 62–71.

12. Nelson, *Birds of Yorkshire*, vol. 2, pp. 774–5.

13. W. R. P. Bourne, 'The past status of the herons in Britain', *Bulletin of the British Ornithological Club*, vol. 119 (1999). Dr Bourne has also drawn attention to the import of live wild fowl (including herons) from the near continent to London in the early sixteenth century. It seems less likely they would be imported earlier as far north as York. See. W. R. P. Bourne, 'Information in the Lisle letters from Calais in the early sixteenth century relating to the development of the English bird trade', *Archives of Natural History*, vol. 26 (1999). See also B. Yapp, *Birds in Medieval Manuscripts* (London, 1981), pp. 108–9; A. H. Evans (ed.), *Turner on Birds* (Cambridge, 1903).

14. Nelson, *Birds of Yorkshire*, vol. 2, p. 775.

15. Ibid., p. 622.

16. Coney, 'Martin Mere'; Sheppard, *Draining of the Marshlands*.

17. J. Mitchell, 'A Scottish bog-hay meadow', *Scottish Wildlife*, vol. 20 (1984), pp. 15–17.

18. H. Stephens, *A Manual of Practical Draining* (London, 1846), pp. 6–11.

19. J. A. Symon, *Scottish Farming, Past and Present* (Edinburgh, 1959), pp. 401–2.

20. Quoted in W. A. Porter, *Tarves Lang Syne: The Story of a Scottish Parish* (York, 1996), p. 36.

21. Stephens, *Practical Draining*, pp. 17–29.

22. J. Smith, *Remarks on Thorough Draining and Deep Ploughing* (Edinburgh, 1831).

23. Symon, *Scottish Farming*, pp. 402–5.

24. H. Stephens, *The Yester Deep Land-Culture* (Edinburgh, 1855), pp. 22, 102–22.

25. T. Allen, *A New and Complete History of the County of York* (London, 1828), vol. 1, pp. 231, 236.

26. C. St John, *Short Sketches of the Wild Sports and Natural History of the Highlands* (London, 1847), p. 168.

27. D. Parker and E. B. Penning-Rowsell, *Water Planning in Britain* (London, 1980), p. 200.

28. *Country Gentlemen's Catalogue* (London, 1894), p. 18.

29. Royal Society for the Protection of Birds, *Wet Grasslands – What Future?* (Sandy, 1993), p. 11.

30. Symon, *Scottish Farming*, p. 409.

31. Scottish Peat and Land Development Association, *Reclamation!* (n.p., n.d.), p. 6.

32. G. M. Binnie, *Early Victorian Water Engineers* (London, 1981), p. 198.

33. S. Patterson, 'The Control of Infectious Diseases in Fife, *c.* 1855–1950', unpublished University of St Andrews PhD thesis, pp. 79–90.

34. Scottish Development Department, *A Measure of Plenty: Water Resources in Scotland, a General Survey* (HMSO, Edinburgh, 1973), p. 47.

35. Quoted in Binnie, *Water Engineers*, pp. 34, 36.

36. Ibid., pp. 55, 180, 201, 266.

37. J. M. Gale, *The Glasgow Water Works, extracted from the Proceedings of the Institute of Engineers in Scotland* (Glasgow, 1864), pp. 10–25.

38. *Glasgow Corporation Water Works, Commemorative Volume* (Glasgow 1877), pp. 4–19.

39. Ibid., p. 20.

40. J. M. Gale, *Report on the Waste of Water* (Glasgow Corporation Water Works, 1860), p. 1; J. M. Gale, *Glasgow Water Works*, pp. 46–7.

41. Quoted in D. Kinnersley, *Troubled Water: Rivers, Politics and Pollution* (London, 1988), p. 45.

42. J. M. Gale, *On the Extension of the Loch Katrine Water Works, reprinted from Transactions of the Institute of Engineers and Shipbuilders of Scotland* (Glasgow, 1895), pp. 1–3.

43. Kinnersley, *Troubled Water*, p. 168.

44. P. L. Payne, *The Hydro: A Study of the Development of the Major Hydro-Electric Schemes Undertaken by the North of Scotland Hydro-Electric Board* (Aberdeen, 1988), p. 6.

45. Ibid., p. 185.

46. Payne, *Hydro*, pp. 214–47; Kinnersley, *Troubled Water*, pp. 89–91.

47. Quoted in E. Porter, *Water Management in England and Wales* (Cambridge, 1978), pp. 25–6.

48. B. W. Clapp, *An Environmental History of Britain since the Industrial Revolution* (London, 1994), p. 80.

49. Ibid., p. 88.

50. J. W. Kempster, *Our Rivers* (Oxford, 1948), pp. 43, 51.

51. J. Sheail, 'Sewering the English suburbs: an inter-war perspective', *Journal of Historical Geography*, vol. 19 (1993), p. 437.

52. Kempster, *Our Rivers*, pp. 54–5.

53. J. Gay et al., 'Environmental implications of the treatment of coastal sewage discharges', in J. C. Currie and A. T. Pepper (eds), *Water and the Environment* (London, 1993), p. 79.

54. J. Sheail, 'Government and the perception of reservoir development in Britain: an historical perspective', *Planning Perspectives*, vol. 1 (1986), p. 54.

55. Clapp, *Environmental History*, pp. 88–95.

56. P. Womack, *Improvement and Romance: Constructing the Myth of the Highlands* (London, 1985), p. 156.

57. N. Hoyle and K. Sankey, *Thirlmere Water: A Hundred Miles, a Hundred Years* (Bury, 1994), pp. 8–19.

58. Porter, *Water Management*, pp. 37–9; J. Sheail, *Nature in Trust: The History of Nature Conservation in Britain* (Glasgow, 1976), pp. 59–60, 83–5.

59. Porter, *Water Management*, pp. 39–42.

60. K. J. Lea, 'Hydro-electric power developments and the landscape in the Highlands of Scotland', *Scottish Geographical Magazine*, vol. 84 (1968), pp. 239–55.

61. *Hansard*, vol. 374 (1940–1), p. 232.

62. *Report of the Committee on Hydro-electric Development in Scotland*, (Cmd. 6406, PP1942–43, IV), p. 34.

63. National Trust for Scotland archives, Hydro Electric Board, Box 1, letter from Stormonth Darling, 9 July 1958.
64. NTS archive, HEB, Box 1, Memorandum to the Mackenzie Committee, 1961.
65. J. Sheail, *Seventy-five Years in Ecology: The British Ecological Society* (Oxford, 1987), pp. 226–31. See also R. Gregory, 'The Cow Green Reservoir', in P. J. Smith (ed.), *The Politics of Physical Resources* (Harmondsworth, 1975).
66. Sheail, 'Government and the perception of reservoir development', p. 57.
67. Porter, *Water Management*, pp. 45–6.
68. K. C. Edwards, H. H. Swinnerton and R. H. Hall, *The Peak District* (London, 1962), pp. 187–8.
69. North of Scotland Hydro-Electric Board, *Highland Water Power* (Edinburgh, n.d. but *c*. 1956), p. 45.
70. Parker and Penning-Rowsell, *Water Resources*, pp. 94–101.

NOTES TO CHAPTER 5

1. W. Pennington, 'Vegetation history in the north-west of England: a regional synthesis', in D. Walker and R. G. West (eds), *Studies in the Vegetational History of the British Isles* (Cambridge, 1970), pp. 72–5.
2. Quoted in W. H. Pearsall, *Mountains and Moorlands* (London, 1950), p. 161.
3. F. Fraser Darling, *Pelican in the Wilderness* (London, 1956), p. 353. This is, as far as I know, his first reference to the Highlands as 'wet desert', and others similarly came to refer to the area as 'semi-desert': W. H. Pearsall, 'Problems of conservation in the Highlands', *Institute of Biology Journal*, vol. 7 (1960), p. 7; W. J. Eggeling in *Natural Resources in Scotland: Symposium at the Royal Society of Edinburgh* (Scottish Council, Development and Industry, 1961), p. 353. Fraser Darling's ideas of the Highlands as degraded had, however, been worked out as early as 1947 in *Natural History in the Highlands and Islands* (London, 1947) and were clearly expressed as soil degradation in *West Highland Survey* (Oxford, 1955), pp. 167–76.
4. Pearsall, *Mountains and Moorland*, p. 188.
5. H. J. B. Birks, 'Long-term ecological change in the British uplands', in M. B. Usher and D. B. A. Thompson (eds), *Ecological Change in the Uplands* (Oxford, 1988), p. 47.
6. R. Tipping, 'The form and fate of Scotland's woodlands', *Proceedings of the Society of Antiquaries of Scotland*, vol. 124 (1994), pp. 26–9.
7. A. F. Brown and I. P. Bainbridge, 'Grouse moors and upland breeding birds', in D. B. A. Thompson, A. J. Hester and M. B. Usher (eds), *Heaths and Moorland: Cultural Landcapes* (Edinburgh, 1995), p. 53.
8. D. A. Ratcliffe and D. B. A. Thompson, 'The British uplands: their ecological character and international significance', in Usher and Thompson (eds), *Ecological Change in the Uplands*, p. 29.
9. G. White, *The Natural History of Selborne*, with introduction by J. E. Chatfield (Exeter, 1981), pp. 144–5.
10. S. Holloway, *The Historical Atlas of Breeding Birds in Britain and Ireland, 1875–1900* (London, 1996), p. 314.
11. White, *Natural History*, pp. 59, 67, 81; D. M. Henderson and J. H. Dickson

(eds), *A Naturalist in the Highlands: James Robertson, His Life and Travels in Scotland, 1767–1771* (Edinburgh, 1994), pp. 155–6; J. C. Atkinson, *Forty Years in a Moorland Parish* (London, 1907), p. 320; D. Raistrick (ed.), *North York Moors* (HMSO, 1969), p. 24.

12. Holloway, *Historical Atlas*, p. 314.

13. Quoted in H. C. Darby, 'Note on the birds of the undrained fen', in D. Lack, *The Birds of Cambridgeshire* (Cambridge, 1934), p. 21.

14. E. V. Baxter and L. J. Rintoul, *The Birds of Scotland: Their History, Distribution and Migration* (Edinburgh, 1953), vol. 2, p. 596.

15. T. H. Nelson, *The Birds of Yorkshire* (London, 1907), vol. 2, pp. 568–9.

16. Thomas Pennant, *A Tour in Scotland*, 1769 (Warrington edn, 1774), p. 115.

17. O. H. Mackenzie, *A Hundred Years in the Highlands* (Edinburgh edn, 1988), p. 104.

18. P. Hudson, *Grouse in Space and Time: The Population Biology of a Managed Gamebird* (Fordingbridge, 1992).

19. D. N. McVean and J. D. Lockie, *Ecology and Land Use in Upland Scotland* (Edinburgh, 1969), p. 41; W. J. Eggeling, 'Nature conservation in Scotland', *Trans. Royal Highland and Agricultural Society*, vol. 8 (1964), pp. 1–27.

20. M. Shoard, 'The lure of the moors', in J. R. Gold and J. Burgess (eds), *Valued Environments* (London, 1982); E. C. Mackey, M. C. Shewry and G. J. Tudor, *Land Cover Change: Scotland from the 1940s to the 1980s* (Edinburgh, 1998), pp. 70–5.

21. The best account is E. Richards, *A History of the Highland Clearances*, 2 vols (London, 1982, 1985).

22. F. Fraser Darling, 'Ecology of land use in the Highlands and Islands', in D. S. Thomson and I. Grimble (eds), *The Future of the Highlands* (London, 1968), p. 38.

23. C. Sydes and G. R. Miller, 'Range management and nature conservation in the British uplands', in Usher and Thompson (eds), *Ecological Change in the Uplands*, p. 332; A. Mather, 'The environmental impact of sheep farming in the Scottish Highlands', in T. C. Smout (ed.), *Scotland Since Prehistory; Natural Change and Human Impact* (Aberdeen, 1993), p. 83.

24. J. D. Milne et al., 'The impact of vertebrate herbivores on the natural heritage of the Scottish uplands – a review', *Scottish Natural Heritage Research Review*, no. 95 (1998).

25. R. H. Marrs, A. Rizand and A. F. Harrison, 'The effect of removing sheep grazing on soil chemistry, above-ground nutrient distribution, and selected aspects of soil fertility in long-term experiments at Moor House National Nature Reserve', *Journal of Applied Ecology*, vol. 26 (1989), pp. 647–61.

26. R. D. Bardgett et al. in *Agricultural Systems and Environment*, vol. 45 (1993), pp. 25–45, cited in Milne et al., 'Impact of vertebrate herbivores', p. 67.

27. See for example, J. Miles, 'Vegetation and soil change in the uplands', and M. B. Usher and S. M. Gardner, 'Animal communities in the uplands: how is naturalness influenced by management?', both in Usher and Thompson (eds), *Ecological Change in the Uplands*, pp. 57–92.

28. Milne et al., 'Impact of vertebrate herbivores', p. 69.

29. Henderson and Dickson (eds), *A Naturalist in the Highlands*, p. 161.

30. J. Macdonald, 'On the agriculture of the counties of Ross and Cromarty',

Transactions of the Highland and Agricultural Society of Scotland, 4th series, vol. 9 (1877), p. 205.

31. J. Hunter, 'Sheep and deer: Highland sheep farming, 1850–1900', *Northern Scotland*, vol. 1 (1973), pp. 203–5.

32. J. Macdonald, 'On the agriculture of the county of Sutherland', *Transactions of the Highland and Agricultural Society of Scotland*, 4th series, vol. 12 (1880), pp. 84–5. See also P. R. Latham, 'The deterioration of mountain pastures and suggestions for the improvement', *Transactions of the Highland and Agricultural Society of Scotland*, 4th series, vol. 15 (1883), p. 112.

33. Sydes and Miller, 'Range management', pp. 326–7.

34. Mather, 'The environmental impact of sheep farming', pp. 64–78.

35. R. Hewson, 'The effect on heather *Calluna vulgaris* of excluding sheep from moorland in north-east England', *The Naturalist*, vol. 102 (1977), pp. 133–6.

36. D. M. McFerran, W. I. Montgomery and J. H. McAdam, 'Effects of grazing intensity on heathland vegetation and ground beetle assemblages of the uplands of Co. Antrim, north-east Ireland', *Proceedings of the Royal Irish Academy*, vol. 94B (1994), pp. 41–52; A. W. Mackay and J. H. Tallis, 'Summit-type blanket mire erosion in the Forest of Bowland, Lancashire, UK: predisposing factors and implications for conservation', *Biological Conservation*, vol. 74 (1996), pp. 31–44; P. A. Tallantine, 'Plant macrofossils from the historical period from Scroat Tarn (Wasdale), English Lake District, in relation to environmental and climatic changes', *Botanical Journal of Scotland*, vol. 49 (1997), pp. 1–17.

37. A. C. Stevenson and D. B. A. Thompson, 'Long-term changes in the extent of heather moorland', *Holocene*, vol. 3 (1993), pp. 70–6.

38. Reay D. G. Clarke, personal communication.

39. Mackenzie, *Hundred Years in the Highlands*, p. 24.

40. A. Mitchell (ed.), *Geographical Collections relating to Scotland made by Walter Macfarlane* (Scottish History Society, Edinburgh, 1906), vol. 3, pp. 271, 276, 281, 300.

41. J. Keay and J. Keay, *Collins Encyclopedia of Scotland* (London, 1994), p. 374.

42. D. E. Allen, *The Naturalist in Britain: A Social History* (London, 1976), pp. 141–2.

43. T. M. Devine, *The Great Highland Famine* (Edinburgh, 1988); Hunter, 'Sheep and deer'.

44. W. Orr, *Deer Forests, Landlords and Crofters* (Edinburgh, 1982), pp. 47–8.

45. Nelson, *Birds of Yorkshire*, vol. 2, pp. 516–17.

46. J. Ruffer, 'Bags of time', *Country Life*, July 30 1987, p. 115.

47. Lord Walsingham and R. Payne-Gallway, *Shooting: Moor and Marsh* (Badminton Library of Sports and Past-times, London, 1889), pp. 36–8; F. Chapman, *Gun, Rod and Rifle* (Eastbourne, 1908), p. 14.

48. G. Scott, *Grouse Land and the Fringe of the Moor* (London, 1937), pp. 40, 72–3.

49. J. Ritchie, *The Influence of Man on Animal Life in Scotland* (Cambridge, 1920), pp. 128–36, 165–7; Pearsall, *Mountains and Moorlands*, pp. 234–6; Baxter and Rintoul, *Birds of Scotland*, vol. 1, p. 309.

50. Scottish Record Office, Breadalbane Muniments, GD 112/16/7/3/24.

51. Cromartie Muniments, GD 305.

52. Scott, *Grouse Land*, p. 56.

53. Walsingham and Payne-Gallway, *Shooting*, p. 101.
54. J. S. Smith, 'Changing deer numbers in the Scottish Highlands since 1780', in Smout (ed.), *Scotland since Prehistory*, pp. 79–88.
55. F. Fraser Darling, *West Highland Survey* (Oxford, 1955), p. 178.
56. See for example, J. A. Baddeley, D. B. A. Thompson and J. A. Lee, 'Regional and historical variation in the nitrogen content of *Racomitrium lanuginosum* in Britain in relation to atmospheric nitrogen deposition', *Environmental Pollution*, vol. 84 (1994), pp. 189–96; N. J. Loader and V. R. Switsur, 'Reconstructing past environmental change using stable isotopes in tree-rings', *Botanical Journal of Scotland*, vol. 48 (1996), pp. 65–78; S. J. Brooks, 'Three thousand years of environmental history in a Cairngorms lochan revealed by analysis of non-biting midges (*Insecta: Diptera: Chironomidae*)', *Botanical Journal of Scotland*, vol. 48 (1996), pp. 89–98; R. W. Battarbee, J. Mason, I. Renberg and J. F. Talling (eds), *Palaeolimnology and Lake Acidification* (Royal Society, London, 1990).
57. R. W. Battarbee et al., 'Palaeolimnological evidence for the atmospheric contamination and acidification of high Cairngorm lochs, with special reference to Lochnagar', *Botanical Journal of Scotland*, vol. 48 (1996), pp. 79–87.
58. Baddeley et al., 'The nitrogen content of *Racomitruim*'.
59. J. A. Lee, J. H. Tallis and S. J. Woodin, 'Acidic deposition and British upland vegetation', in Usher and Thompson (eds), *Ecological Change in the Uplands*, pp. 151–62.
60. M. G. R. Cannell, D. Fowler and C. E. R. Pitcairn, 'Climate change and pollutant impacts on Scottish vegetation', *Botanical Journal of Scotland*, vol. 49 (1997), pp. 301–13.
61. J. T. de Smidt, 'The imminent destruction of northwest European heaths due to atmospheric nitrogen deposition', in Thompson, Hester and Usher (eds), *Heaths and Moorland*, pp. 206–17; S. E. Hartley, 'The effects of grazing and nutrient inputs on grass-heather competition', *Botanical Journal of Scotland*, vol. 49 (1997), pp. 315–24.
62. R. E. Green, 'Long-term declines in the thickness of eggshells of thrushes, *Turdus* spp., in Britain', *Proceedings of the Royal Society*, series B, vol. 265 (1998), pp. 679–84; S. J. Langan, A. Lilley and B. F. L. Smith, 'The distribution of heather moorland and the sensitivity of associated soils to acidification', in Thompson, Hester and Usher (eds), *Heaths and Moorland*, pp. 218–23.

NOTES TO CHAPTER 6

1. From surveys by the Scottish Tourist Board as reported to the SNH conference in Aberdeen, September 1998; Scottish Natural Heritage, *Access to the Countryside for Open-air Recreation* (Battleby, 1998), p. 21.
2. M. Nicholson, *The Environmental Revolution: A Guide for the New Masters of the World* (London, 1970), p. 44.
3. J. Gaze, *Figures in a Landscape: A History of the National Trust* (n.p., 1988), pp. 10–11.
4. R. Williams, *The Country and the City* (London, 1973).
5. P. Skipworth, *The Great Bird Illustrators and their Art, 1730–1930* (London, 1979), pp. 24–5.

6. *A Memoir of Thomas Bewick Written by Himself*, ed. I. Bain (Oxford, 1979), pp. 8, 204–5.

7. Ibid., pp. 170, 172.

8. Ibid., pp. 52, 65.

9. H. Taylor, *A Claim on the Countryside: A History of the British Outdoor Movement* (Edinburgh, 1997); also, T. Stephenson, *Forbidden Land: The Struggle for Access to Mountain and Moorland* (Manchester, 1989).

10. Taylor, *Claim*, pp. 58–9.

11. Ibid., pp. 20, 23, 30.

12. Ibid., p. 21.

13. Ibid., pp. 21–7.

14. Ibid., p. 30.

15. Scottish Record Office: GD 335, Records of the Scottish Rights of Way Society.

16. J. S. Mill, *Principles of Political Economy*, ed. W. J. Ashley (1909), p. 235.

17. *Scots Magazine*, vol. 6 (1890), no. 31, pp. 1–6; SRO: GD 335. See also T. C. Smout and R. A. Lambert (eds), *Rothiemurchus: Nature and People on a Highland Estate, 1500–2000* (Edinburgh, 1999).

18. Taylor, *Claim*, pp. 129–30; Stephenson, *Forbidden Land*, pp. 123–6.

19. Taylor, *Claim*, p. 122; Stephenson, *Forbidden Land*, pp. 130–8.

20. Taylor, *Claim*, pp. 134–5; D. E. Allen, *The Naturalist in Britain: A Social History* (London, 1976), pp. 164, 170.

21. R. Aitken, 'Stravagers and marauders', *Scottish Mountaineering Club Journal*, vol. 30 (1975), pp. 351–8; Taylor, *Claim*, pp. 131–3.

22. P. Smith, 'Access to mountains', *Blackwoods Magazine*, vol. 150 (1891), p. 265.

23. *Scottish Ski Club Magazine*, vol. 1 (1909), p. 5.

24. 'Report of the committee on non-church going', *Reports on the Schemes of the Church of Scotland* (Edinburgh, 1890–6).

25. D. Macleod, *Dumbarton, Vale of Leven and Loch Lomond: Historical, Legendary, Industrial and Descriptive* (Dunbarton, n.d.), pp. 184–5.

26. Quoted in T. Stephenson, *Forbidden Land*, p. 136.

27. Smith, 'Access to the mountains', pp. 259–72.

28. Stephenson, *Forbidden Land*, pp. 153–64.

29. Gaze, *Figures in a Landscape*, pp. 33–4.

30. H. D. Rawnsley, 'Footpath preservation, a national need', in *Contemporary Review* (September 1886), p. 385.

31. P. Gibb, 'The landowner's error', *Agenda*, no. 3 (autumn 1998), p. 15.

32. G. K. Chesterton, in *Architect's Journal*, 15 August 1928.

33. C. Williams-Ellis, *England and the Octopus* (Portmeirion edn, 1975), pp. 1, 16, 127, 143, 162.

34. C. Williams-Ellis (ed.), *Britain and the Beast* (London, 1937), p. 64.

35. York University: Borthwick Institute, Women's Institute Records, C.2.

36. B. L. Thompson, *The Lake District and the National Trust* (Kendal, 1946), p. 13.

37. W. Birtles and R. Stein, *Planning and Environmental Law* (London, 1994); G. E. Cherry, *Environmental Planning 1939–1969*, vol. 2, *National Parks and Recreation in the Countryside* (London, HMSO, 1975).

38. Quotes in Cherry, *Environmental Planning*, pp. 34–5, 63, 105.

39. See W. M. Adams, *Nature's Place: Conservation Sites and Countryside Change*

(London, 1986), pp. 68–9.

40. Scottish Record Office: Forestry Commission (Scotland) records, FC 9/1–4.

41. *Newsletter of the National Trust for Scotland*, no. 2 (1949); no. 14 (1956); no. 18 (1958). Quotation is from the last, p. 14.

42. *National Parks and the Conservation of Nature in Scotland*, Cmd. 7235 (HMSO, 1947), p. 5.

43. SRO: FC 9/2 – 'Report by Mr Robert Grieve on a visit to Switzerland'.

44. *National Parks and the Conservation of Nature*, pp. 41ff.

45. *Newsletter*, no. 2 (1949), p. 7.

46. *Scotland's Natural Heritage*, no. 14 (1999), p. 5.

47. J. Sheail, 'John Dower, national parks, and town and country planning in Britain', *Planning Perspectives*, vol. 10 (1995), pp. 1–16; M. Shoard, 'The lure of the moors', in J. R. Gold and J. Burgess, *Valued Environments* (London, 1982), pp. 55–73.

48. *Report on National Parks in England and Wales*, Cmd. 6628 (HMSO, 1945), pp. 15, 21.

49. Cherry, *Environmental Planning*; A. and M. McEwen, *National Parks: Conservation or Cosmetics?* (London, 1982).

50. J. Sheail, 'From aspiration to implementation – the establishment of the first National Nature Reserves in Britain', *Landscape Research*, vol. 21 (1996), pp. 37–54. See also E. M. Nicolson, *Britain's Nature Reserves* (London, 1957); P. Marren, *England's National Nature Reserves* (London, 1994).

51. Adams, *Nature's Place*, p. 83.

52. CLA and NFU, joint statement over Exmoor, 1967.

53. D. A. Bigham, *The Law and Administration Relating to Protection of the Environment* (London, 1994), pp. 57, 62.

54. A. Samstag, *For Love of Birds: The Story of the Royal Society for the Protection of Birds 1889–1988* (London, 1988); P. Lowe and J. Goyder, *Environmental Groups in Politics* (London, 1983), pp. 37, 140, 157; Thompson, *The Lake District*, p. 65.

55. Lowe and Goyder, *Environmental Groups*, p. 28.

56. Ibid., p. 59.

57. Quoted in J. Sheail, 'Government and the perception of reservoir development in Britain: an historical perspective', *Planning Perspectives*, vol. 1 (1986), p. 45.

58. H. Newby, *Green and Pleasant Land? Social Change in Rural England* (Harmondsworth, 1980), pp. 255–6.

59. Lowe and Goyder, *Environmental Groups*, p. 10.

60. Adams, *Nature's Place*, p. 76.

61. Newby, *Green and Pleasant*, p. 218.

62. Adams, *Nature's Place*, p. 88.

63. Ibid., ch. 4.

64. Ibid., p. 188.

65. Newby, *Green and Pleasant*, pp. 218–19.

66. Scottish Office, *People and Nature: A New Approach to SSSI Designations in Scotland* (n.p., n.d.), p. 7.

SELECT BIBLIOGRAPHY

Note: A few items dated 1998 or 1999 are highly useful but have appeared since the text was prepared. They are marked with an asterisk.

Adams, W. M. (1986) *Nature's Place: Conservation Sites and Countryside Change*, London.

Adams, W. M. (1997) 'Rationalization and conservation: ecology and the management of nature in the United Kingdom', *Transactions of the Institute of British Geographers*, vol. 22, pp. 277–91.

Aitken, R. (1975) 'Stravagers and marauders', *Scottish Mountaineering Club Journal*, vol. 30, pp. 351–8.

Allen, D. E. (1976) *The Naturalist in Britain: A Social History*, London.

Baddeley, J. A., Thompson, D. B. A. and Lee, J. A. (1994) 'Regional and historical variation in the nitrogen content of *Racomitrium languinosum* in Britain in relation to atmospheric nitrogen deposition', *Environmental Pollution*, vol. 84 (1994) pp. 65–78.

Bailey, R. G., José, P. V. and Sherwood, B. R. (eds) (1998) *United Kingdom Floodplains Management* (Westbury, 1998).

Barber, S. (1998) 'The history of the Coniston woodlands, Cumbria, UK', in K. J. Kirby and C. Watkins (eds), *The Ecological History of European Forests*, Wallingford, pp. 167–84.

Bate, J. (1991) *Romantic Ecology: Wordsworth and the Environmental Tradition*, London.

Battarbee, R. W. et al. (1996) 'Palaeolimnological evidence for the atmospheric contamination and acidification of high Cairngorm lochs, with special reference to Lochnagar', *Botanical Journal of Scotland*, vol. 48, pp. 79–87.

Baxter, E. V. and Rintoul, L. J. (1953) *The Birds of Scotland: Their History, Distribution and Migration*, 2 vols, Edinburgh.

Berry, R. J. and Johnston, J. L. (1980) *The Natural History of Shetland*, London.

Bewick, T. (1979) *A Memoir Written by Himself*, ed. I. Bain, Oxford.

Bigham, D. A. (1994) *The Law and Administration Relating to Protection of the Environment, London*.

Binnie, G. M. (1981) *Early Victorian Water Engineers*, London.

Birks, H. J. B. (1988) 'Long-term change in the British uplands', in M. B. Usher and D. B. A. Thompson, *Ecological Change in the Uplands*, Oxford, pp. 37–56.

Birtles, W. and Stein, R. (1994) *Planning and Environmental Law*, London.

Blaxter, K. and Robertson, N. (1995) *From Dearth to Plenty: the Modern Revolution in Food Production*, Cambridge.

Boserup, E. (1965) *The Conditions of Agricultural Growth: the Economics of Agrarian Change under Population Pressure*, London.

191

Bourne, W. R. P. B. (1999) 'The past status of herons in Britain', *Bulletin of the British Ornithological Club*, vol. 119.

Breeze, D. (1998) 'The Great Myth of Caledon', in T. C. Smout (ed.) *Scottish Woodland History*, Edinburgh, pp. 47–51.

Brooks, S. F. (1996) 'Three thousand years of environmental history in a Cairngorms lochan revealed by analysis of non-biting midges', *Botanical Journal of Scotland*, vol. 48, pp. 89–98.

Brown, A. F. and Bainbridge, I. P. (1995) 'Grouse moors and upland breeding birds', in D. B. A. Thompson, A. J. Hester and M. B. Usher (eds), *Heaths and Moorland: Cultural Landscapes*, Edinburgh, pp. 51–66.

Campbell, B. E. S. and Overton, M. (eds) (1991) *Land, Labour and Livestock: Historical Studies in European Agricultural Productivity*, Manchester.

Cannell, M. G. R., Fowler, D. and Pitcairn, C. E. R. (1997) 'Climate change and pollutant impacts on Scottish vegetation', *Botanical Journal of Scotland*, vol. 49, pp. 301–13.

Cheape, H. (1993) 'Woodlands on the Clanranald estates: a case study', *Scotland Since Prehistory: Natural Change and Human Impact*, Aberdeen.

Cherry, G. E. (1975) *Environmental Planning 1939–1969*, vol. 2, *National Parks and Recreation in the Countryside*, London.

Clapp, B. W. (1994) *An Environmental History of Britain since the Industrial Revolution*, London.

Clark, G. (1992) 'The economics of exhaustion, the Postan thesis and the Agricultural Revolution', *Journal of Economic History*, vol. 52 (1992) pp. 61–84.

Coney, A. (1992) 'Fish, fowl and fen: landscape economy in seventeenth-century Martin Mere', *Landscape History*, vol. 14, pp. 51–64.

Cook, H. J. (1996) 'Physicians and natural history', in Jardine, Secord and Spary (eds) *Cultures of Natural History*, Cambridge, pp. 91–105.

Currie, J. C. and Pepper, A. T. (eds, 1993) *Water and the Environment*, London.

Darling, F. F. (1947) *Natural History in the Highlands and Islands*, London; revised later as Darling, F. F. and Boyd, J. M. (1964) *The Highlands and Islands*, London.

Darling, F. F. (1955) *West Highland Survey*, Oxford.

Darling, F. F. (1956) *Pelican in the Wilderness*, London.

Darling, F. F. (1968) 'Ecology of land use in the Highlands and Islands', in D. S. Thomson and I. Grimble (eds) *The Future of the Highlands*, London, pp. 27–56.

Davidson, D. A. and Simpson, I. A. (1994) 'Soils and landscape history: case studies from the Northern Isles of Scotland', in S. Foster and T. C. Smout, *The History of Soils and Field Systems*, Aberdeen, pp. 66–74.

Davidson, D. A. and Smout, T. C. (1996) 'Soil change in Scotland', in A. G. Taylor, I. E. Gordon and M. B. Usher, *Soil, Sustainability and the Natural Heritage*, Edinburgh, pp. 44–54.

de Smidt, J. T. (1995) 'The imminent destruction of northwest European heaths due to atmospheric nitrogen deposition', in D. B. A. Thompson, A. J. Hester and M. B. Usher (eds) *Heaths and Moorland: Cultural Landscapes*, Edinburgh.

Devine, T. M. (1988) *The Great Highland Famine*, Edinburgh.

Devine, T. M. (1994) *The Transformation of Rural Scotland: Social Change and the Agrarian Economy, 1660–1815*, Edinburgh.

Dickson, J. H. (1993) 'Scottish woodlands: their ancient past and precarious future', *Scottish Forestry*, vol. 47, pp. 73–8.

Dingwall, C. (1997) 'Coppice management in Highland Perthshire', in T. C. Smout (ed.) *Scottish Woodland History*, Edinburgh, pp. 162–75.

Dodgshon, R. A. (1978) 'Land improvement in Scottish farming: marl and lime in Roxburghshire and Berwickshire in the eighteenth century', *Agricultural History Review*, vol. 26, pp. 11–14.

Dodgshon, R. A. (1981) *Land and Society in Early Scotland*, Oxford.

Dodgshon, R. A. (1993) 'Strategies of farming in the western Highlands and Islands prior to crofting and the clearances', *Economic History Review*, vol. 46, pp. 679–701.

Dodgshon, R. A. (1994) 'Budgeting for survival: nutrient flow and traditional Highland farming', in S. Foster and T. C. Smout (eds), *History of Soils and Field Systems*, pp. 83–93.

Dodgshon, R. A. and Olsson, E. G. (1988) 'Productivity and nutrient use in eighteenth–century Scottish Highland townships', *Geografiska Annaler*, vol. 70B, pp. 39–51.

Dunlop, B. M. S. (1997) 'The woods of Strathspey in the nineteenth and twentieth centuries', in T. C. Smout (ed.) *Scottish Woodland History*, Edinburgh, pp. 176–89.

Edwards, K. C., Swinnerton, H. H. and Hall, R. H. (1962) *The Peak District*, London.

Eggeling, W. J. (1964) 'Nature conservation in Scotland', *Transactions of the Royal Highland and Agricultural Society*, vol. 8, pp. 1–27.

Fleming, A. (1997) 'Towards a history of wood pasture in Swaledale (North Yorkshire)', *Landscape History*, vol. 19, pp. 57–74.

Foster, S. and Smout, T. C. (eds) (1994) *The History of Soils and Field Systems*, Aberdeen.

Garritt, J. (1998) 'Politics, knowledge, action: the local implementation of the Convention on Biological Diversity', unpublished conference paper, University of Lancaster.

Gaze, J. (1988) *Figures in a Landscape: A History of the National Trust*, n.p.

Gilvear, D. J. and Winterbottom, S. J. (1998) 'Changes in channel morphology, floodplain land use and flood damage on the rivers Tay and Tummell over the last 250 years: implications for floodplain management', in R. G. Bailey, P. V. José and B. R. Sherwood (eds) *United Kingdom Floodplains*, London, pp. 93–116.

Gold, J. R. and Burgess, J. (eds) (1982) *Valued Environments*, London.

Green, R. E. (1998) 'Long-term declines in the thickness of eggshells of thrushes', *Proceedings of the Royal Society*, series B, vol. 265, pp. 679–84.

Grove, J. (1988) *The Little Ice Age*, London.

Gulliver, R. (1998) 'What were woods like in the seventeenth century? Examples from the Helmsley Estate, North-east Yorkshire', in K. J. Kirby and C. Watkins (eds), *The Ecological History of European Forests*, Wallingford, pp. 135–54.

Harrison, B. (1973) 'Animals and the state in nineteenth-century England', *English Historical Review*, vol. 349, pp. 786–820.

Hartley, S. E. (1997) 'The effects of grazing and nutrient inputs on grass-heather competiton', *Botanical Journal of Scotland*, vol. 49, pp. 315–24.

*Hassan, J. (1998) *A History of Water in Modern England and Wales*, Manchester.

Henderson, D. M. and Dickson, J. H. (eds) (1994) *A Naturalist in the Highlands:*

James Robertson, His Life and Travels in Scotland, 1767–1771, Edinburgh.

Hewson, R. (1977) 'The effect on heather *Calluna vulgais* of excluding sheep from moorland in north-east England', *The Naturalist*, vol. 102, pp. 133–6.

Holdgate, M. W. (1997) 'Standards, sustainability and integrated land use', *Macaulay Land Use Research Institute 10th Anniversary Lectures, Aberdeen.*

Holloway, S. (1996) *The Historical Atlas of Breeding Birds in Britain and Ireland, 1875–1900*, London.

Hoyle, N. and Sankey, K. (1994) *Thirlmere Water: A Hundred Miles, a Hundred Years*, Bury.

Hudson, P. (1992) *Grouse in Space and Time: The Population Biology of a Managed Gamebird*, Fordingbridge.

Hunter, J. (1973) 'Sheep and deer: Highland sheep farming, 1850–1900', *Northern Scotland*, vol. 1 (1973) pp. 199–222.

Hunter, J. (1995) *On the Other Side of Sorrow: Nature and People in the Scottish Highlands*, Edinburgh.

Hunter, J. M. (1986) *Land into Landscape*, Harlow.

Jardine, N., Secord, J. A. and Spary, E. C. (eds) (1996) *Cultures of Natural History*, Cambridge.

Jenkins, D. (ed.) (1985) *Biology and Management of the River Dee*, Institute of Terrestrial Ecology, Abbots Ripton.

Jenkins, D. (ed.) (1988) *Land Use in the River Spey Catchment*, Aberdeen Centre for Land Use, Aberdeen.

Jones, M. (1998) 'The rise, decline and extinction of spring wood management in south-west Yorkshire', in C. Watkins (ed.), *European Woods and Forests: Studies in Cultural History*, Wallingford, pp. 55–72.

Kempster, J. W. (1948) *Our Rivers*, Oxford.

Kinnersley, D. (1988) *Troubled Water: Rivers, Politics and Pollution*, London.

Kirby, K. J. and Watkins, C. (eds) (1998) *The Ecological History of European Forests*, Wallingford.

Kitchener, A. C. (1998) 'Extinctions, introductions and colonisations of Scottish mammals and birds since the last Ice Age', in R. Lambert (ed.), *Species History in Scotland: Introductions and Extinctions since the Ice Age*, Edinburgh, pp. 63–92.

Kjærgaard, T. (1994) *The Danish Revolution, 1500–1800: An Ecohistorical Interpretation*, Cambridge.

Koerner, L. (1996) 'Carl Linnaeus in his time and place', in N. Jardine, J. A. Secord and E. C. Spary (eds), *Cultures of Natural History*, Cambridge, pp. 145–62.

Lambert, R. (1998) 'From exploitation to extinction, to environmental icon: our images of the great auk', in R. Lambert (ed.), *Species History in Scotland: Introductions and Extinctions since the Ice Age*, Edinburgh, pp. 20–37.

Lambert, R. (ed.) (1998) *Species History in Scotland: Introductions and Extinctions since the Ice Age*, Edinburgh.

Langan, S. J., Lilley, A. and Smith, B. F. L. (1995) 'The distribution of heather moorland and the sensitivity of associated soils to acidification', in D. B. A. Thompson, A. J. Hester and M. B. Usher (eds), *Heaths and Moorland: Cultural Landscapes*, Edinburgh.

Lasdun, S. (1991) *The English Park: Royal, Private and Public*, London.

Lea, K. J. (1941) 'Hydro-electric power developments and the landscape in the

Highlands of Scotland', *Scottish Geographical Magazine*, vol. 84, pp. 239–55.

Lee, J. A., Tallis, J. H. and Woodin, S. J. (1988) 'Acidic deposition and British upland vegetation', in M. B. Usher and D. B. A. Thompson (eds), *Ecological Change in the Uplands*, Oxford, pp. 151–62.

Lindsay, J. M. (1974) 'The use of Woodland in Argyllshire and Perthshire between 1650 and 1850', unpublished University of Edinburgh thesis.

Lindsay, J. M. (1975) 'Charcoal iron smelting and its fuel supply: the example of Lorn Furnace, Argyllshire, 1753–1876', *Journal of Historical Geography*, vol. 1, pp. 283–98.

Loader, N. J. and Switsur, V. R. (1996) 'Reconstructing past environmental change using stable isotopes in tree-rings', *Botanical Journal of Scotland*, vol. 48, pp. 65–78.

Lowe, P. and Goyder, J. (1983) *Environmental Groups in Politics*, London.

Mabey, R. (1980) *The Common Ground*, London.

McFerran, D. M., Montgomery, W. I. and McAdam, J. H. (1994) 'Effects of grazing intensity on heathland vegetation and ground beetle assemblages of the uplands of Co. Antrim, north-east Ireland', *Proceedings of the Royal Irish Academy*, vol. 94B, pp. 41–52.

Mackay, A. W. and Tallis, J. H. (1996) 'Summit-type blanket mire erosion in the Forest of Bowland, Lancashire, UK: predisposing factors and implications for conservation', *Biological Conservation*, vol. 74, pp. 31–44.

Mackenzie, O. H. (1988) *A Hundred Years in the Highlands*, Edinburgh.

Mackey, E. C., Shewry, M. C. and Tudor, G. J. (1998) *Land Cover Change: Scotland from the 1940s to the 1980s*, Edinburgh.

Macleod, A. (ed. and trans.) (1952) *The Songs of Duncan Ban Macintyre*, Scottish Gaelic Text Society, Edinburgh.

McVean, D. N. and Lockie, J. D. (1969) *Ecology and Land Use in Upland Scotland*, Edinburgh.

Marchant, J. H. et al. (1990) *Population Trends in British Breeding Birds*, Thetford.

Marren, P. (1994) *England's National Nature Reserves*, London.

Marrs, R. H., Rizand, A. and Harrison, A. F. (1989) 'The effect of removing sheep grazing on soil chemistry, above-ground nutrient distribution, and selected aspects of soil fertility in long-term experiments at Moor House National Nature Reserve', *Journal of Applied Ecology*, vol. 26, pp. 647–61.

Mather, A. (1993) 'The environmental impact of sheep farming in the Scottish Highlands, in T. C. Smout (ed.) *Scotland Since Prehistory*, Aberdeen, pp. 79–88.

Mellanby, K. (1981) *Farming and Wildlife*, London.

Mercer, R. and Tipping, R. (1994) 'The prehistory of soil erosion in the northern and eastern Cheviot hills, Anglo-Scottish Borders', in S. Foster and T. C. Smout (eds) *The History of Soil and Field Systems*, Aberdeen, pp. 1–25.

Miles, J. (1988) 'Vegetation and soil change in the uplands', in M. B. Usher and D. B. A. Thompson (eds) *Ecological Change in the Uplands*, Oxford, pp. 57–70.

Miles, J. (1994) 'The soil resource and problems today', in S. Foster and T. C. Smout (eds) *Soils and Field Systems*, Aberdeen, pp. 145–58.

Milne, J. D. et al. (1998) 'The impact of vertebrate herbivores on the natural heritage of the Scottish uplands', *Scottish Natural Heritage Research Review*, no. 95.

Mitchell, A. (ed.) (1906) *Geographical Collections Relating to Scotland, Made by*

Walter Macfarlane, 3 vols, Scottish History Society, Edinburgh.

Mitchell, I. (1998) *Scotland's Mountains before the Mountaineers*, Edinburgh.

Mitchell, J. (1984) 'A Scottish bog-hay meadow', *Scottish Wildlife*, vol. 20, pp. 15–17.

Monk, S. (1935) *The Sublime: A Study of Critical Theories in Eighteenth Century England*, New York.

Moore, N. W. (1987) *The Bird of Time: The Science and Politics of Nature Conservation*, Cambridge.

Murray, W. H. (1962) *Highland Landscape, a Survey*, Edinburgh.

Nairne, D. (1892) 'Notes on Highland woods, ancient and modern', *Transactions of the Gaelic Society of Inverness*, vol. 17, pp. 170–221.

Nelson, T. H. (1907) *The Birds of Yorkshire: A Historical Account of the Avifauna of the County*, 2 vols, London.

Nethersole-Thompson, D. and Watson, A. (1981) *The Cairngorms: Their Natural History and Scenery*, Perth.

Newby, H. (1980) *Green and Pleasant Land? Social Change in Rural England*, Harmondsworth.

Newman, E. I. and Harvey, P. D. A. (1997) 'Did soil fertility decline in medieval English farms? Evidence from Cuxham, Oxfordshire, 1320–1340', *Agricultural History Review*, vol. 45, pp. 119–36.

Nicholson, E. M. (1957) *Britain's Nature Reserves*, London.

Nicholson, M. (1970) *The Environmental and Revolution: A Guide for the New Masters of the World*, London.

Nicholson, M. H. (1959) *Mountain Gloom and Mountain Glory: the Development of the Aesthetics of the Infinite*, New York.

Noyes, R. (1973) *Wordsworth and the Art of Landscape*, New York.

O'Connor, R. J. and Shrubb, M. (1986) *Farming and Birds*, Cambridge.

Olwig, K. (1984) *Nature's Ideological Landscape*, London.

Orr, W. (1982) *Deer Forests, Landlords and Crofters*, Edinburgh.

Parker, D. and Penning-Rowsell, E. G. (1980) *Water Planning in Britain*, London.

Payne, P. J. (1988) *The Hydro: A Study of the Development of the Major Hydro-Electric Schemes Undertaken by the North of Scotland Hydro-Electric Board*, Aberdeen.

Pearsall, W. H. (1950) *Mountains and Moorlands*, London.

Pennington, W. (1970) 'Vegetation history in the north-west of England: a regional synthesis', in D. Walker and R. G. West (eds), *Studies in the Vegetational History of the British Isles*, Cambridge, pp. 41–80.

Porter, E. (1978) *Water Management in England and Wales*, Cambridge.

Pretty, J. (1990) 'Sustainable agriculture in the Middle Ages: the English manor', *Agricultural History Review*, vol. 38, pp. 1–19.

Postan, M. M. (1973) *Essays in Medieval Agriculture and General Problems of the Medieval Economy*, Cambridge.

Radkau, (1997) 'The wordy worship of nature and the tacit feeling for nature in the history of German forestry', in M. Teich, R. Porter and B. Gustafsson (eds), *Nature and Society in Historical Context*, Cambridge, pp. 228–39.

Ratcliffe, D. A. and Thompson, D. B. A. (1988) 'The British uplands: their ecological character and international significance', in M. B. Usher and D. B. A. Thompson (eds), *Ecological Change in the Uplands*, Oxford, pp. 9–36.

Richards, E. (1982–5) *A History of the Highland Clearances*, 2 vols, London.

Ritchie, J. (1920) *The Influence of Man on Animal Life in Scotland*, Cambridge.

Ryle, G. (1969) *Forest Service: The First Forty-five Years of the Forestry Commission of Great Britain*, Newton Abbot.

Samstag, A. (1988) *For Love of Birds: The Story of the Royal Society for the Protection of Birds*, London.

Schama, S. (1995) *Landscape and Memory*, London.

Scott, G. (1937) *Grouse Land and the Fringe of the Moor*, London.

Scottish Council (Development and Industry) (1961) *Natural Resources in Scotland: Symposium at the Royal Society of Edinburgh*, Edinburgh.

Scottish Natural Heritage (1995) *The Natural Heritage of Scotland: An Overview*, Perth.

Shaw, J. (1994) 'Manuring and fertilising the Scottish Lowlands', in Foster and T. C. Smout (eds), *The History of Soils and Field Systems*, Aberdeen, pp. 111–18.

Sheail, J. (1976) *Nature in Trust: The History of Nature Conservation in Britain*, Glasgow.

Sheail, J. (1981) *Rural Conservation in Interwar Britain*, Oxford.

Sheail, J. (1985) *Pesticides and Nature Conservation: The British Experience, 1950–1975*, Oxford.

Sheail, J. (1986) 'Government and the perception of reservoir development in Britain: an historical perspective', *Planning Perspectives*, vol. 1, pp. 45–60.

Sheail, J. (1987) *Seventy-five Years in Ecology: The British Ecological Society*, Oxford.

Sheail, J. (1993) 'Sewering the English suburbs: an inter-war perspective', *Journal of Historical Geography*, vol. 19, pp. 433–47.

Sheail, J. (1995) 'Elements of sustainable agriculture: the UK experience, 1840–1940', *Agricultural History Review*, vol. 43, pp. 178–92.

Sheail, J. (1995) 'John Dower, national parks, and town and country planning in Britain', *Planning Perspectives*, vol. 10, pp. 55–73.

Sheail, J. (1996) 'From aspiration to implementation – the establishment of the first National Nature Reserves in Britain', *Landscape Research*, vol. 21, pp. 37–54.

*Sheail, J. (1998) *Nature Conservation in Britain – The Formative Years*, London.

Sheail, J. (1998) *Regional Distribution of Wealth in England as Indicated in the 1524–5 Lay Subsidy Returns*, List and Index Society Special Series, vol. 28.

Sheppard, J. A. (1958) *The Draining of the Hull Valley*, East Yorkshire Local History Society.

Sheppard, J. A. (1966) *The Draining of the Marshlands of South Holderness*, East Yorkshire Local History Society.

Shiel, R. S. (1991) 'Improving soil productivity in the pre-fertiliser era', in B. E. S. Campbell and M. Overton (eds), *Land, Labour and Livestock: Historical Studies in European Agricultural Productivity*, Manchester, pp. 51–77.

Shoard, M. (1980) *Theft of the Countryside*, London.

Shoard, M. (1982) 'The lure of the moors', in J. R. Gold and J. Burgess (eds), *Valued Environments*, London, pp. 55–74.

Shoard, M. (1987) *This Land is Our Land: Struggle for Britain's Countryside*, London.

Sinclair, J. (ed.) (1972–83) *Statistical Accounts of Scotland, 1791–1799*, new edition by I. R. Grant and D. J. Withrington, Wakefield.

Smith, P. J. (1975) *The Politics of Physical Resources*, Harmondsworth.

Smith, W. (ed.) (1883) *Old Yorkshire*.

Smout, T. C. (ed.) (1993) *Scotland since Prehistory: Natural Change and Human Impact*, Aberdeen.

Smout, T. C. (1994) 'Trees as historic landscapes: Wallace's oak to Reforesting Scotland', *Scottish Forestry*, vol. 48, pp. 244–52.

Smout, T.C. (1997) 'Cutting into the pine: Loch Arkaig and Rothiemurchus in the eighteenth century', in T. C. Smout (ed.), *Scottish Woodland History*, Edinburgh, pp. 115–25.

Smout, T. C. (ed.) (1997) *Scottish Woodland History*, Edinburgh.

Smout, T.C. (1999) 'The history of the Rothiemurchus woodlands', in T. C. Smout and R. Lambert (eds), *Rothiemurchus: Nature and People on a Highland Estate, 1500–2000*, Edinburgh, pp. 60–78.

Smout, T. C. and Foster, S. (eds) (1994) *The History of Soil and Field Systems*, Aberdeen.

Smout, T.C. and Watson, F. (1997) 'Exploiting semi-natural woods, 1600–1800', in T. C. Smout (ed.), *Scottish Woodland History*, Edinburgh, pp. 86–100.

Smout, T.C. and Lambert, R. (eds) (1999) *Rothiemurchus: Nature and People on a Highland Estate, 1500–2000*, Edinburgh.

Stamp, L. D. (1964) *Man and the Land*, London.

Stephenson, T. (1989) *Forbidden Land: The Struggle for Access to Mountain and Moorland*, Manchester.

Stevens, A. C. and Thompson, D. B. A. (1993) 'Long-term changes in the extent of heather moorland', *Holocene*, vol. 3, pp. 70–6.

Stone, J. C. (1989) *The Pont Manuscript Maps of Scotland: Sixteenth-century Origins of a Blaeu Atlas*, Tring.

Sydes, C. and Miller, G. R. (1988) 'Range management and nature conservation in the British uplands', in M. B. Usher and D. B. A. Thompson (eds), *Ecological Change in the Uplands*, Oxford, pp. 323–38.

Symon, J. A. (1959) *Scottish Farming Past and Present*, Edinburgh.

Tallantine, P. A. (1997) 'Plant macrofossils from the historical period from Scroat Tarn (Wasdale) English Lake District, in relation to environmental and climatic changes', *Botanical Journal of Scotland*, vol. 49, pp. 1–17.

Taylor, A. G., Gordon, J. E. and Usher, M. B. (eds) (1996) *Soil, Sustainability and the Natural Heritage*, Edinburgh.

Taylor, H. (1997) *A Claim on the Countryside: A History of the British Outdoor Movement*, Edinburgh.

Teich, M., Porter, R. and Gustafsson, B. (eds, 1997) *Nature and Society in Historical Context*, Cambridge.

Thomas, K. (1983) *Man and the Natural World: Changing Attitudes in England, 1500–1800*, London.

Thompson, B. L. (1946) *The Lake District and the National Trust*, Kendal.

Thompson, D. B. A., Hester, A. J. and Usher, M. B. (eds) (1995) *Heaths and Moorland: Cultural Landscapes*, Edinburgh.

Thomson, D. (1995) *An Introduction to Gaelic Poetry*, London.

Thomson, D. S. (ed.) (1994) *The Companion to Gaelic Scotland*, Oxford.

Thomson, D. S. and Grimble, I. (eds) (1968) *The Future of the Highlands*, London.

Tipping, R. (1994) 'The form and fate of Scotland's woodlands', *Proceedings of the Society of Antiquaries of Scotland*, vol. 124, pp. 1–54.

Tivy, J. (ed.) (1973) *The Organic Resources of Scotland: Their Nature and Evaluation*, Edinburgh.

Tsouvalis-Gerber, J. (1998) 'Making the invisible visible: ancient woodlands, British forest policy and the social construction of reality', in C. Watkins (ed.), *European Woods and Forests: Studies in Cultural History*, Wallingford, pp. 215–30.

Usher, M. B. (1996) 'The soil ecosystem and sustainability', in A. G. Taylor, J. E. Gordon and M. B. Usher (eds), *Soil Sustainability and the Natural Heritage*, Edinburgh, pp. 22–41.

Usher, M. B. and Gardner, S. M. (1988) 'Animal communities in the uplands: how is naturalness influenced by management?', in M. B. Usher and D. B. A. Thompson (eds), *Ecological Change in the Uplands*, Oxford, pp. 75–92.

Usher, M. B. and Thompson, D. B. A. (eds) (1988) *Ecological Change in the Uplands*, Oxford.

Veitch, J. (1887) *Feeling for Nature in Scottish Poetry*, 2 vols, Edinburgh.

Walker, D. and West, R. G. (eds) (1970) *Studies in the Vegetational History of the British Isles*, Cambridge.

Walker, G. J. and Kirby, K. J. (1989) *Inventories of Ancient, Long-established and Semi-natural Woodland for Scotland*, Nature Conservancy Council Research and Survey in Nature Conservation, no. 22.

Walsingham, Lord and Payne-Galloway, R. (1889) *Shooting: Moor and Marsh*, Badminton Library of Sports and Past-times, London.

Watkins, C. (ed.) (1998) *European Woods and Forests: Studies in Cultural History*, Wallingford.

Watson, A. (1983) 'Eighteenth-century deer numbers and pine regeneration near Braemar, Scotland', *Biological Conservation*, vol. 25, pp. 289–305.

Watson, F. (1997) 'Sustaining a myth: the Irish in the West Highlands', *Scottish Woodland History Discussion Group Notes*, 2, Institute for Environmental History, University of St Andrews.

Watson, F. (1997) 'Rights and responsibilities: wood management as seen through baron court records', in T. C. Smout (ed.), *Scottish Woodland History*, Edinburgh, pp. 101–14.

Watson, F. (1998) 'Need versus greed? Attitudes to woodland management on a central Scottish highland estate, 1630–1740', in C. Watkins (ed.), *European Woods and Forests: Studies in Cultural History*, Wallingford, pp. 135–56.

Williams, R. (1973) *The Country and the City*, London.

Williams-Ellis, C. (ed.) (1937) *Britain and the Beast*, London.

Williams-Ellis, C. (1975) *England and the Octopus*, Portmeirion.

Winchester, A. J. L. (1987) *Landscape and Society in Medieval Cumbria*, Edinburgh.

Woodell, S. R. J. (ed.) (1985) *The English Landscape, Past, Present and Future*, Oxford.

Woodward, D. (1994) '"Gooding the earth": manuring practices in Britain, 1500–1800', in S. Foster and T. C. Smout (eds), *The History of Soils and Field Systems*, Aberdeen, pp. 101–10.

Womack, P. (1985) *Improvement and Romance: Constructing the Myth of the Highlands*, London.

Wyatt, J. (1995) *Wordsworth and the Geologists*, Cambridge.
*Yalden, D. (1999) *The History of British Mammals*, London.
Yapp, B. (1981) *Birds in Medieval Manuscripts*, London.

INDEX